D0881664

# About Island Press

Since 1984, the nonprofit Island Press has been stimulating, shaping, and communicating the ideas that are essential for solving environmental problems worldwide. With more than 800 titles in print and some 40 new releases each year, we are the nation's leading publisher on environmental issues. We identify innovative thinkers and emerging trends in the environmental field. We work with world-renowned experts and authors to develop cross-disciplinary solutions to environmental challenges.

Island Press designs and implements coordinated book publication campaigns in order to communicate our critical messages in print, in person, and online using the latest technologies, programs, and the media. Our goal: to reach targeted audiences—scientists, policymakers, environmental advocates, the media, and concerned citizens—who can and will take action to protect the plants and animals that enrich our world, the ecosystems we need to survive, the water we drink, and the air we breathe.

Island Press gratefully acknowledges the support of its work by the Agua Fund, Inc., The Margaret A. Cargill Foundation, Betsy and Jesse Fink Foundation, The William and Flora Hewlett Foundation, The Kresge Foundation, The Forrest and Frances Lattner Foundation, The Andrew W. Mellon Foundation, The Curtis and Edith Munson Foundation, The Overbrook Foundation, The David and Lucile Packard Foundation, The Summit Foundation, Trust for Architectural Easements, The Winslow Foundation, and other generous donors.

# Observation and Ecology

# Observation and Ecology

## BROADENING THE SCOPE OF SCIENCE TO UNDERSTAND A COMPLEX WORLD

RAFE SAGARIN AND ANÍBAL PAUCHARD

Library of Congress Cataloging-in-Publication Data

Sagarin, Rafe.
Observation and ecology : broadening the scope of science to understand a complex world / Rafe Sagarin and Anibal Pauchard.
p. cm.
Includes bibliographical references and index.
ISBN 978-1-59726-825-7 (cloth : acid-free paper) — ISBN 1-59726-825-9 (cloth : acid-free paper) — ISBN 978-1-59726-826-4 (paper : acid-free paper) — ISBN 1-59726-826-7 (paper : acid-free paper)
1. Ecology—Methodology. 2. Ecology—Research. 3. Ecology—Study and teaching. 4. Observation (Scientific method). 5. Nature observation. 6. Biocomplexity. I. Pauchard, Anibal. II. Title.
QH541.28.S23 2012
577.072—dc23 2012001570

Printed on recycled, acid-free paper

Manufactured in the United States of America

10 9 8 7 6 5 4 3 2 1

*Keywords:* Island Press, observational approaches to ecology, natural history, conservation biology, ecological science, citizen science, traditional ecological knowledge, local ecological knowledge, climate change, complex systems, social ecological science, education and policy in ecology.

CONTENTS

## BOXES

BOXES

# FOREWORD

Paul K. Dayton

> The best of science doesn't consist of mathematical models and experiments, as textbooks make it seem. Those come later. It springs fresh from a more primitive mode of thought, wherein the hunter's mind weaves ideas from old facts and fresh metaphors and the scrambled crazy images of things recently seen. To move forward is to concoct new patterns of thought, which in turn dictate the design of the models and experiments. Easy to say, difficult to achieve.
>
> — E. O. Wilson, *The Diversity of Life* (1992)

Skillful observations are the foundation of ecological science. Ecosystems are complex and confusing. They are composed of a large number of species and a vast number of interactions in which the relationships are nonlinear and characterized by many thresholds. Because the best approaches are not always obvious, we rely on classical analytical techniques of simplification, analysis, and synthesis. Each step depends absolutely on the good observations of sound natural history. The process of simplifying nature is difficult but essential. Science's main objective is to make interesting, accurate generalizations about nature based on as few relevant parameters as necessary—all nature is trivially related, and marginally important parameters need to be weeded out so as to focus on the parameters essential for the generalization. Discovering the appropriate simplification is a critical component of the scientific process.

Our objective is to understand processes, not only patterns that we study with observations. One relies on astute observations by skilled naturalists to define hypotheses about the processes that seem to be particularly important, and these hypotheses are tested in many legitimate ways that depend on correct understanding of nature. Ideally, models are created to generalize the processes that have emerged from the tests. Each step relies on observations of nature, and a good ecologist must have a broad synthetic mind, an ability to practice strong inference, and a sense of place or a feel for nature (that is, they must be respectful, alert, observant, and intuitive).

These procedures are well understood, but many have lamented the fact that powerful molecular and analytic tools have been joined with general theoretical models that are not actually grounded in nature. While such theoretical approaches can be very powerful, real understanding and solutions to environmental problems must rely on life history biology, a deep understanding of taxonomy, the identification of strong interactions, and an intuitive appreciation of complex ecosystem dynamics. Unfortunately, this latter expertise has fallen from favor in academe. A common problem is that hypotheses are based on and tested with inappropriate simplifications and assumptions due to a lack of good natural history. The danger is that bad assumptions can be measurable and precise, esthetically pleasing and apparently useful, but the hypotheses may be irrelevant to the natural world and/or make the right predictions for the wrong reasons.

Real progress in understanding nature must be based, first, on a deep sensitivity to natural patterns and processes across broad scales in space and time, and, further, on a deep understanding of the literature and the many powerful tools needed to test and generalize the results of scientific investigations. Our success as scientists depends upon successful integration of general theory and natural history. This book is dedicated to the goal of recovering a respect for excellent observations of nature. Such observations are fundamental to every component of the process of doing meaningful ecological research. This book needs to be read and appreciated by ecologists in particular and all biology instructors in general.

# ACKNOWLEDGMENTS

We both are extremely grateful for the thoughtful contributions made by all of the authors of the featured boxes. Our editor at Island Press, Barbara Dean, displayed keen insight, careful attention, and immense patience during every stage of the production of this book. Paul Alaback, Martin Nuñez, Brendon Larson, Christoph Kueffer, and members of the Laboratorio de Invasiones Biológicas (LIB) discussed ideas about the book with us and provided critical feedback. Retta Breugger, Benjamin DeGain, Ami Kidder, Laura Marsh, Kristin Wisneski, students in Rafe's first seminar on observational life sciences at the University of Arizona, provided great examples and critical feedback on many of the ideas that made it into this book.

Rafe appreciates the support of a John Simon Guggenheim Memorial Foundation Fellowship, which was instrumental during the writing of the book. The Institute of the Environment and the Office of the Vice Provost for Research at the University of Arizona also supported the research and writing of this book. The Stanford University Libraries Department of Special Collections repeatedly granted access to the Edward Ricketts papers, which are referred to throughout this book. Rafe's ideas about the value of observational approaches have been shaped by two scientific mentors, Chuck Baxter, lecturer emeritus at Stanford University's Hopkins Marine Station, and Steven Gaines, Dean of the Bren School at the University of California, Santa Barbara. Chuck helped Rafe learn how to see the natural world in a scientific and philosophical way, and, as his graduate advisor, Steve taught Rafe that good questions and critical analyses can make good science out of all kinds of observations. Finally, Rafe would like to thank his wife Rebecca Crocker and daughters, Ella and Rosa, for their patience, good humor, and high spirits throughout the writing of this book and throughout his meandering, sometimes frustrating, and often adventurous career as an observational ecologist.

Aníbal would like to acknowledge Paula Díaz for her endless patience during the writing of this book and for sharing ideas about the connection between psychology and ecology. Aníbal would like to dedicate this book to his son, Benjamín, who has been a boundless and always surprising source of inspiration in thinking about ecology and society. Aníbal would like to thank Paul Alaback, who has provided him with an excellent example of a naturalist, ecologist, and scientist. Aníbal was partially funded by the Institute of Ecology and Biodiversity through the grants CONICYT Basal Funding grant PFB-23 and Iniciativa Científica Milenio of the Chilean Ministry of Economy, grant P05-002. Special thanks to the MECESUP grant and the Universidad de Concepción, which funded Aníbal's stay at the University of Arizona.

INTRODUCTION

# A Time of Change and Adaptation in Ecology

All of us are living in a time of transformation—economic, social, political, and environmental changes are challenging us everywhere and constantly. It seems obvious, then, that the science of ecology, which deals with the tangled web of relationships between organisms and the biogeochemical world we live in, should also be in a transformative period. The methods, goals, participants, and even philosophies of ecology are changing. The changes we are seeing now are wrought from a convergence of unprecedented environmental challenges and remarkable new opportunities to study ecological systems. Both the signal of this change in ecological science and the vehicle for ongoing transformation is how we use *observation* to discover new phenomena, to achieve ecological understanding, and to share ecological ideas.

This book is about harnessing the power of observation in order to participate in this most unique time to study ecology. In its essence, all ecology is primarily about the observation of nature, but in the reality of academic ecology, observations are quickly transformed into theory that gets tested on a computer, or experimental treatments in the field or in a lab that are manipulated to test well-defined hypotheses. These are important ways of achieving ecological understanding, methods that have dominated ecology for the last half century or so, but they have limitations that become apparent the more ecological systems change.

In this book we focus on "observation-based ecology," which we define as the ecology that relies on observations of systems that have not been manipulated for scientific purposes. This is a broad definition that subsumes a wide range of powerful ways of observing and making sense of ecological systems. Uncovering these approaches, their promise and their pitfalls, is what this book is about.

What do these kinds of observations look like? They may be the field notes of naturalists on the timing of spring blooms, logs from old whaling ships documenting the extent of pack ice, or very-high-resolution satellite images that compile data on primary productivity. They may concern the most basic ecological questions like "Why do starfish come in different colors?" or the most pressing socio-ecological concerns like "How fast will avian influenza spread?" They may be focused intently on one particular protein and what it says about an animal's environment, or they may be sweeping views of interacting population, nutrient, and temperature cycles through decades of time and across entire continents. These observations may be the products of long-term government-sponsored monitoring programs, or the collective efforts of citizens who count birds in their neighborhood every Christmas, or the stories of old fishermen who have been meticulously documenting changes to their fishing grounds over decades. In other words, the observations that are becoming critical to today's ecological understanding can come from anywhere and anyone, even if they have been taken with a completely different, often nonscientific, purpose in mind. They can deal with the most minute as well as the largest scales of space and biology, they can be snapshots of single moments in Earth history or long series of observations made through decades, centuries, or millennia.

How do we use these observations? Some observational approaches simply provide new discoveries about ecological systems and thus are not much different from the approach of early naturalist-explorers. Oftentimes, though, we are building an understanding of ecological dynamics by correlating one set of observed data against another, like looking at the relationship between long-term increases in temperature and long-term advancement of springtime events like budding and migration. Sometimes we are checking observations against an expected pattern, as for example, when we look at the geographic ranges of butterflies through time to see if they meet the expectation that as climate warms, species' ranges will shift toward the cooler poles. And given that we are living in a changing planet, observations are valuable for their ability to document these changes, especially in recent decades as humans' footprints on Earth become ever harder to ignore. The varying concentrations of ozone in the upper atmosphere over the southern pole, patterns of El Niño oscillations, and the spread of an invasive agricultural pest across a landscape are all examples of uncontrolled transformations that can be studied observationally.

Are the kinds of observations we are talking about "scientific"? Observations taken as part of an experiment generally don't receive this scrutiny because we tend to think of experimentation and manipulation of data as being at the heart of what science is and what scientists do. Such an observation only exists because a scientist devised a way to test a hypothesis and then recorded what resulted. But the kinds of observations we're discussing in this book can be put into a scientific context before or after—sometimes long after—they are made.

Because of the often unplanned nature of how these observations are made, great care must be taken in their interpretation. Usually, no scientist prescreened them or planned in advance exactly what observations should be made. Even in the case of long-term data from a monitoring program designed by scientists, the observations are often ultimately used in unexpected ways.

In many cases, the kinds of observations we discuss here are also uncontrolled—many variables, like predation, climatic factors, population size, and maybe human impacts are interacting over large scales of time and space in observational data sets. Often we didn't get to choose which of these variables made it into the data set and which were controlled out of it. In some cases, though, there are "natural" controls that can be used to test the effects of a variable of interest (Diamond and Robinson 2010). For example, the now-restricted area around the Chernobyl nuclear plant in Ukraine, which exploded in 1986, is a control of sorts on the impacts of humans on wildlife. Without humans present for 25 years we can now observe a proliferation of wildlife and even the selective evolution of organisms absent human impacts (Mycio 2006).

In most cases, though, where controlled comparisons aren't inherent in the data, it is up to the ecologist to make controls after the fact by dividing the data in ways that isolate different factors. For example, when Rafe studied changes to the tide pool communities of Monterey Bay between 1930 and 1993, he obviously couldn't control factors like water quality during the intervening six decades, but he could look at how filter feeders (which would be more affected by changes in water quality than other animals) fared relative to scavengers or predators.

Despite options for dealing with unmanipulated and uncontrolled variables, observation-based ecology still raises the question "What is science?"—and underneath that question lie other uncertainties that make some scientists nervous. How can the musings of an old dead naturalist be trusted? How can observations from a long-retired whaling fleet be replicated? What can imagery from miles above Earth tell us about the mechanisms of ecological interactions down on the ground? These specific types of questions reflect a more general criticism of observational approaches. This line of criticism emphasizes that we can't get at the mechanisms underlying ecological phenomena just by observing them. Or that observations of naturalists and fishermen are just anecdotal "just-so stories" that may sound interesting, but don't amount to hard evidence, thus confusing rather than solving ecological questions. By this view, observations amount to "stamp collecting"—a hobby without a greater purpose. And there is always the admonishment, heard many times in critiques of our own work, that "correlation does not imply causation." All of these arguments have legitimate roots, and none of them can be dismissed with a single blanket defense—in other words, they must be asked of every observational study every time. At the same time, none of these arguments are fatal to the premise that observational approaches, even without experimental manipulations, can be a legitimate source of scientific ecological understanding.

In many cases, observational approaches may be the only way to understand some ecological phenomena, especially as those phenomena grow in scale or become more inseparable, in their causes and effects, from human activity. We argue in this book that our innate observational skills are enormously powerful and underutilized. We believe that these skills can be trained to be better, and that even when our sharply honed observational skills reach their limits, we can extend them still further, expanding the scale and resolution of our observations, by fusing our innate senses with new technologies. We also reject the notion that our observational skills are too prone to bias to be trusted. In fact, we argue the opposite—that through the process of becoming more astute observers of environmental change at multiple scales of space and time we become more aware of our potential biases and thus better able to account for them.

In recent decades, there has been an increasing amount of ecological research that relies mainly on observational data. The shift in ecology toward embracing observational methods is neither speculative nor is it a passing trend—both quantitatively and qualitatively it is very real. There are already discernible trends in scientific publishing that indicate this shift; for example, three leading peer-reviewed journals now publish greater percentages of primarily observational studies than they did 20 years ago, as we discuss in Chapter 2. New observational data sets and long-term monitoring schemes are arising despite poor economic conditions.

But a lot of the change in ecology is not easily quantified. This is in part because it is happening at the level of students who are eager to take an expansive view of both the methods and outcomes of their research, but whose contributions are underrepresented in the overall "phenotype," or outward appearance, of ecological science. Students are sometimes spreading their interest in observational methods by dragging their advisors along into their "new" ways of doing ecology. But these students are not alone. Even ecologists who have built entire careers on cleverly designed experimental approaches are rediscovering and advocating the power of observation. Furthermore, the need to rapidly address global problems is forcing ecologists to jump into the unpredictable waters of observation-based ecology.

Taken together all this means that there is momentum behind the expansion of observational approaches in ecology, and we predict that they will continue to play a larger role in ecology. But there are two unusual characteristics of this growth. First, we don't think this growth must come at the expense of other approaches in ecology. An increased appreciation of the power of observational studies doesn't mean they will supplant experimental or theoretical approaches. In fact just the opposite will occur: embracing observation makes experimental methods more valuable and more efficient because instead of trying to answer questions that are better addressed through unmanipulated observations, experiments can be used strategically to fill in the gaps that remain once many observations have been thrown at a problem. Observational approaches also make theory more valuable because they provide a pathway for validating theory with real data. We see this growth of observation as something that will be fully integrated within ecology. There is no need to declare a new field of ecology in the way that "conservation biology" or "molecular ecology" have their own courses and journals and specialized language and professional societies. Rather, the most important power of observational approaches is their potential to create better integration within ecological science and between ecological science and the larger world.

Observation-based ecology will bring more people and ideas into the tent of ecology, both because of its simplicity and its complexity. The simplicity of going out into nature and counting, measuring, watching, and recording opens ecology to a non-elite, nonprofessional world where people who don't spend their lives as ecologists can nonetheless contribute to ecological science as collectors of data and as consumers of ecological ideas that they can then spread and inculcate into other endeavors, like politics and art. The complex side of observation-based ecology is a challenge that is already being answered by all kinds of scientists who may not consider themselves ecologists, from statisticians with new approaches to handling data, to molecular biologists with new techniques for observing ecological relationships at the smallest scales, to space scientists who are designing new missions to asteroids to study the likely ecology of the early Earth.

## What You Will Find in This Book

We address four main challenges in this book. First, we want to paint a clear picture of what observational approaches to ecology are and where they fit in the context of the changing nature of ecological science. Second, we want to consider the full range of observational abilities we have available to us—from our innate observational capacities (which go way beyond what we can see) to our technologies, and to the many keen observers of the world who do not even consider themselves to be scientists or ecologists. Third, we want to consider the challenges and practical difficulties of using a primarily observational approach to achieve a scientific understanding of the ecological world. Finally, we want to show how observations can be a bridge from ecological science to education, environmental policy, and resource management. The book is thus divided into four parts reflecting these challenges.

Part I sets up the framework for understanding the role of observation-based ecology as part of a scientific and societal enterprise. In Chapter 1 we lay out what observation-based ecology looks like, how it relates to its ancestry in natural history and why it is different from the dominant experimental mode of doing ecology. We illustrate the scope of observational approaches, using examples from our own work and others to demonstrate that observational approaches can be used across a wide range of activities related to ecological science. We will outline the sometimes surprising range of data sources, from Alaskan gambling contests to centuries-old church records, that have already contributed to our modern understanding of ecological change. In Chapter 2 we trace the cyclical history that observational approaches have had within formal ecological science, from the late nineteenth century when, for example, naturalist Teddy Roosevelt complained about his studies at Harvard that "the tendency was to treat as not serious, as unscientific, any kind of work that was not carried on with laborious minuteness in the laboratory" (Millard 2006), to the recent revival of interest in natural history as a valid mode of scientific inquiry (e.g., Greene 2005; Dayton 2003). This chapter will take us to the present junction, when both traditional natural history and new observational approaches that would have been foreign to early natural historians are assuming a more respectable role within today's ecological sciences.

Part II covers the "how" of observation-based ecology. We start in Chapter 3 by illustrating the importance of utilizing multiple observational senses to achieve ecological understanding. Using examples such as paleoecologist Geerat Vermeij's remarkable observational abilities despite his lifelong blindness, we show that abundant ecological information exists beyond our visual field. In Chapter 4 we expand our observational capabilities even further by reviewing the wide array of new technologies that allow us to expand our innate observational senses into previously unfathomable expanses of space, time, and sensory spectra. Remote sensing, for instance, allows us to view phenological changes and waves of species invasions across entire regions and at different spatial scales (Pauchard and Shea 2006). Molecular biology, which in the twentieth century caused a deep rift between naturalists and supposedly more "rigorous" biologists (Wilson 1994), lends itself to observational approaches that are now being fully integrated with ecological studies (e.g., Kozak, Graham, and Wiens 2008; Alter, Rynes, and Palumbi 2007; Sagarin and Somero 2006). Animal-borne sensors are essentially turning animals into observers of the natural world and in the process overturning longstanding assumptions about the basic ecology of even well-studied organisms (Moll et al. 2007; Block 2005). Here we consider as a metaphor ecologist Carlos Martinez del Rio's imagined "ecological cyborg" (Martinez del Rio 2009)—an organism that combines the observational skills of a scientist, the passion of a naturalist and the technical acumen of a robot. In Chapter 5 we argue that the resurgence of observational approaches presents an unprecedented opportunity for creating a more inclusive ecological science, a trend that is becoming evident in the much greater deference now paid to local and traditional forms of ecological knowledge, in the emergence of citizen-science programs that are both an educational tool and a rich data source, and, critically, in an embracing of social-science methodologies.

Despite our occasional unbridled exuberance, this book is meant to be a primer about the promise and pitfalls of an expanded role of observation in ecology. Therefore, in Part III we will take pains to address the known and potential unknown shortcomings of observational approaches. In Chapter 6 we deal with the practical questions that arise: How can we deal with the deluge of data that often comes with observational approaches? Or on the other hand, how can we deal with spotty data sets, often collected by observers long ago? How can we do observation-based ecology under adverse circumstances, especially in the developing world where ecology is both resource- and data-poor? Such questions arise to some degree in all science, but they are especially acute when there are significant constraints to the scale of ecological studies or to the kinds of generalizations that can be drawn from them. We will argue in this book that we can no longer count on being able to manipulate all the variables we would like to if we are going to advance in ecology, but doing science without manipulation resurrects many questions that drove ecology toward experimentalism in the first place. In Chapter 7, we will tackle these challenges in the same way we experienced them as observational ecologists—as difficult questions that arose during the course of planning our research, or in long, doubt-filled hours in the field, or as the critiques we received as we began to share our work (and were grilled, in our early presentations, by advisors, graduate committee members, and ruthless professors). When are correlations between data strong enough to be scientifically defensible? Can we do science without clear hypotheses? And how can we uncover the underlying mechanisms of ecological interactions when all we have is what has been observed?

Part IV focuses on what can be done with all these observations and how they can have a real impact on our society, going beyond the traditional avenues of academic publishing and conference presentations. Here again, we return to the potential power of observation-based studies to affect education, policy, and management related to natural resources and environmental change. In Chapter 8 we argue that observational approaches can be especially influential and informative to environmental policy debates. We consider that observational approaches don't just convey the technical information needed in order to make sound, clear policy decisions, but they also influence the emotional and sociological aspects of policy making in a way that few other types of science can. In Chapter 9, we argue that the same visceral reactions that may place observational studies in the center of policy debates also make observation-based studies suitable for public education by enhancing our capacities to relate to nature and its environmental and conservation problems. Both in the collection and analysis of data from nature, and in the presentation of observation-based ecological studies, observational approaches naturally translate into compelling narratives and metaphors that can be communicated in a range of media. This includes formal and informal science education, from reinvigorating simple natural history–based field courses to nature films that use animal-borne sensors to reveal the ecology of organisms that few have had access to previously.

We close by considering some of the emergent properties of an observational approach to ecology. How can embracing observational approaches get us away from our labs and computers and back to the appreciation of nature that launched most of our careers in science? And likewise, how can society get closer to nature by embracing some basic principles of observation-based ecology?

Unlike a textbook, this book will mix the rather sober analysis of issues like science philosophy with very personal reflections of our enthusiasm for observational studies and the struggles we've had working in a manner that is still not fully part of the mainstream of scientific ecology. Moreover, although we use quantitative analyses where possible, many of the take-home messages we hope to impart will be told through stories, which we argue (in Chapter 7) are valid media for expressing scientific thoughts. Some of these stories will come from ecologists themselves, in the form of text boxes written by some of the most creative ecologists we know. We hope, then, that you won't find reading this book to be a chore—also unlike a textbook.

The spirit of this book is that there are compelling and important questions emerging across vast stretches of space and time on our continually changing planet and that an increasing reliance on observation-based ecology—a trend that is already occurring—may allow us to finally answer these questions. We hope that our treatment of sometimes-controversial topics will breach the old barriers between experimental and observational approaches. We recognize that experimental manipulations have played and will always continue to play a vital role in ecology. Just as there are questions experiments can't answer, so too are there questions that simple observations will never be able to parse into useful components. Indeed, we argue that the strongest ecological studies will combine observational and experimental approaches in an iterative back-and-forth exchange between ways of achieving ecological understanding.

The most important lesson from this book, we hope, is that it is an incredibly *exciting* time to be involved in the science of ecology. We are in the midst of a new era of discovery. Advances in observational technologies have documented new species and even whole phyla of organisms (Bourlat et al. 2006) and have revealed surprising new discoveries about species as familiar as squirrels (Rundus et al. 2007), as valued as bluefin tuna (Block et al. 2005), and as revered as great whales (Alter, Rynes, and Palumbi 2007). The discoveries being made are both astounding and frightening. Moreover, the increasing involvement of nonscientists in ecological observation (see Chapter 5) is erasing the boundary between scientist and nonscientist. And observational approaches afford a more straightforward transition between science in practice and science in the public eye, dissolving the perceived boundary between scientific communication and public communication. Observation-based ecology is built out of stories that emerge directly from observations; when newspaper reporters and school children and filmmakers and public officials ask us "What's going on with this system?" we can begin to answer them directly—not with a mumbled, heavily qualified explanation contingent on a dubious scaling-up of results from a square meter to thousands of square kilometers, but with a straightforward call to simply look at the data. And finally, the boundaries defining what can be studied in scientific ecology are also falling away. Even the most staid and traditional university ecology departments are now finding their ranks filled with students and young professors looking toward economics, law, public policy, history, and anthropology to guide their inquiries. An unexpected side effect of this is that today's ecologists, especially students, are liberated from the pressure of "finding publishable results" from experiments having a fairly narrow scope. Rather, as an ever-larger set of variables becomes available for examination, the opportunities for serendipitous discovery of something completely unexpected are as great as they were when Darwin sailed around the world on the *Beagle*.

Just as in those early days of ecological discovery, now there are deeply compelling reasons to embrace observation-based ecology. The planet is changing—in many cases as a result of our failure to steward natural systems—but we now have an enhanced ability to understand the patterns and magnitude of those changes and share that knowledge with people all over the world.

PART I

# THE ROLE OF OBSERVATION IN ECOLOGICAL SCIENCE

Ecology has always been a science based on observing the natural world, so what has changed that we should now draw our attention to the act of observation within a scientific context? In a word, everything. Making sense of just how profound are the changes to our environment and our way of studying the environment requires putting ecology into its historical context. The opening part of this book looks at ecology's ancient roots in natural history, its more modern manifestations as a rigorous science that has become well established in academic institutions, and its current trajectory toward becoming a multidisciplinary science, one that is more fully integrated with societal activities and concerns.

In Chapter 1, we lay out the premise that ecology has always been an adaptable science and argue that while a primarily experimental ecology has served us well in the twentieth century, present conditions are pushing ecology to adapt to a new niche where broad observations are an increasingly important means, sometimes the only means, of making sense of a complex world. In Chapter 2 we dig a little deeper into the history of observations in ecology, to trace the evolution of ecology and to lay out why observations are now more prominent and more powerful than ever before.

CHAPTER 1

# An Observational Approach to Ecology

To understand how ecology will serve us in this era age of rapid environmental change, we need to understand that ecology is not a static discipline. It is continuously *adapting* to the changing world that ecologists find themselves living and working within. This chapter is about the most recent adaptation in ecology, which can be seen in both an increased use and increased diversity of observational approaches to understanding ecological phenomena. This adaptation, like stepwise adaptations in nature, hasn't created an entirely new and unrecognizable entity, but rather has grown recursively from the past state of ecology. Accordingly, we first discuss what ecology was for much of its existence and then we explore how the urgency of environmental change and the opportunity to study that change in unprecedented ways is providing a pathway for adaptation of the science of ecology.

## Ecology as an Experimental Science

One of the prominent characteristics of the ecology since the mid-twentieth century has been the importance of experimental methods. This itself was an evolution from previous ecological methods. During this time, ecology left behind its exploratory stage and progressed through stepwise advances by means of cleverly designed and carefully controlled planned experiments at relatively small scales in order to isolate the mechanisms underlying various ecological phenomena. This is an attractive way to do science. By setting up experiments that tweak just a small number of variables and strict controls, one can often determine with some certainty that a particular causal factor leads to a particular ecological change. For example, an experiment to look at the effects of predation would be set up by erecting barriers around a plot to keep predators out, and the control would be plots where predators roam freely, and there may also be controls on the experimental equipment such as partial barriers that let predators in while allowing the researcher to determine whether the experimental equipment itself might have affected outcomes through shading or the disturbance of installing the equipment.

The manipulative experimental approach is also amenable to replication, provided there is enough space to place multiple copies of the experimental and control plots. This gives a researcher confidence that she can test a *hypothesis*—that is, a testable supposition, such as, "diversity of species in this grassland is maintained by herbivory on species *x*, which would otherwise overgrow all the other species"—about an ecological phenomenon. If the system is amenable to experimental treatment, a good experimental ecologist will probably be able to conceive of not just one, but multiple alternative hypotheses to test. Testing multiple alternative hypotheses that could be serially rejected was the aspiration of John Platt's hugely influential "Strong Inference" (Platt 1964) approach to ecology. In the early 1960s, Platt argued that ecology as a science would forever remain a second-tier endeavor relative to apparently nobler scientific pursuits like chemistry, physics, and molecular biology, until it got its act together and developed a more rigorous framework.

It is easy to see why this experimental approach has been so widely adopted by ecologists. With a manipulative experiment, you know you are going to get a result, or you know the steps you need to take to get a result. Well, at least you are more likely than not to get a result—in reality many experiments go awry because of unexpected forces of nature (maybe strong El Niño storms that rip all your experimental plots off the intertidal rocks where the plots were painstakingly installed). And while experimental work is rarely easy—our colleagues have spent countless hours scuba diving in frigid Alaskan waters, trekking up South Pacific highlands in 99 percent humidity, and mucking about in malarial swamps to deploy, check, repair, and reap data from their experimental setups—it is fairly tractable. That is, it is very likely that someone could conceive of, plan, deploy, analyze, and write about a good experiment within the duration of an extended field course or a graduate-school career. And most important, these features make experimental work inherently fundable, because the experiment has a specific purpose, clear methodological stages, and a relatively constrained set of possible outcomes—very little is left to chance. Once the experiment has been conceived, it is fairly straightforward to explain to a funding agency like the National Science Foundation (NSF) that the experiment will perform as promised, that it will deliver a particular set of data, and that it will answer a particular set of ecological questions.

Manipulative experiments and strong inference have long been important in ecology. They have been used to tackle questions across the spectrum of ecological inquiry—from what controls the dynamics of an ecological community, to why does that animal behave in such an odd way, to how does a limpet navigate its way home to the same spot after every high tide? At the same time, it is easy to see why we as ecologists have been forced to expand outward from this niche. Manipulated experiments can only do so much. And it happens that where they fall short is exactly in the place where we now desperately need more ecological understanding. The scale and the dynamics of many observed ecological phenomena have leapt beyond the scales of time and space that are readily controlled in experiments. In particular, the really big environmental problems we face today—global climate change, collapsing biodiversity, ocean acidification, nitrification of huge water bodies, and the widespread emergence of invasive species and new infectious diseases, to name a few—are all very difficult to study by manipulating variables and repeating cleverly designed experiments.

You can certainly put some marine creatures into a beaker of seawater, drop the pH a few points and see if they can still form calcified shells, and that is important knowledge. But it's going to tell you precious little about the fate of those same creatures spread out across an entire ocean basin that is acidifying due to carbon deposition in some places but not others as the organisms navigate its swirling eddies and trash gyres, experience countless ecological interactions, and evolve with the constantly changing conditions. In other words, both the *scale* and the *dynamics* of small laboratory and field experiments often bear little resemblance to what is going on in the larger world. And then, even if we could get the funding and could work out the logistics of experimentally testing and controlling for all these complex dynamics at the scales at which they work, would it be ethical to do so? It doesn't seem to make sense, if we are worried about the potentially catastrophic effects of large-scale environmental change like ocean acidification, to replicate these changes on a grand experimental scale.

There is also an urgency to the environmental problems we are facing that places an enormous burden on ecological studies. In order to be truly useful for both identifying and potentially solving these problems, we need information quickly (as in *now*), we need it to tell us about what is going on across large spatial scales, and we need it to tell us something about the relationship between the human social and nonhuman ecological components at the heart of the problem. These things are way outside the niche of typical experimental ecological studies.

## Adapting to Change

But even as ecology is outgrowing its niche, it is already adapting to deal with these difficulties. What does this adaptation in ecology look like? We argue in this book that it is based in observational approaches and that it may look like a return to the old ways of ecology, but it is also a lot more than that. For example, there is a strong element of good old-fashioned natural history—the ancient human practice of observing and recording the diversity and changes of nature (as Tom Fleischner helps illuminate for us in Box 1.1)—in the new observational approaches we are seeing. There are, in fact, many concepts of what "natural history" is (Attenborough 2007; Fleischner 2005; Arnold 2003; Dayton and Sala 2001; Applegate 1999; Bartholomew 1997), and undoubtedly there will be times in this book where our ideas converge almost fully with one of them, and there will be times where we diverge quite far from the usual definitions of natural history. Our concept of observational approaches to ecology is both more and less than natural history. It is more than natural history because it incorporates remote observations, like those from satellite mapping and cameras strapped to whales, that are far removed from the human experience of nature usually associated with natural history (although some bold thinkers like Carlos Martinez del Rio argue that modern natural historians should fully embrace these technologies as part of their practice, see Chapter 4). Ecology is also less than natural history because we are, as much as possible, limiting our discussion to the scientific practice of ecology, whereas natural history, although potentially scientific, also widely embraces writing and poetry and art and philosophy. (See naturalhistory network.org for examples of the broad scope of natural history.)

The observational approaches to ecology we discuss in this book also reflect a return to earlier ecological inquiries because they are often integrative of the social component of ecological systems, both in the types of data they are using and the types of questions they are addressing. Early ecologists were naturalists who took painstaking observations of natural systems and attempted to piece those observations together into a more holistic understanding of the world. Many were devoted to the idea that by understanding ecological systems we could gain understanding of human social systems. They were also surprisingly *interdisciplinary* without ever invoking that awkward word. Working after the horrors of the First World War and in the growing shadow of the Second, they were intensely interested in what studies of the relationships of organisms in nature had to say about conflict and cooperation among humans. Warder Allee, for example, felt that unexpected benefits came from cooperation among animals and that similar emergent benefits could accrue to human societies that modeled themselves after animal communities (Allee 1951, 1943). One of his students, the marine ecologist Edward Ricketts, noted that "the laws of animals must be the laws of men" and further refined his thinking through fruitful collaborations with writers like John Steinbeck and philosophers such as the mythologist Joseph Campbell (Rodger 2006; Tamm 2004).

BOX 1.1

### Natural History: The Taproot of Ecology

THOMAS L. FLEISCHNER

A dozen college students lean into the steep hillside above the snout of the enormous valley glacier. For the moment, though, they pay no heed to the massive muscle of ice—their attention is focused, laser-like, on the enchanting internal structures within tubular corollas. The world suddenly takes on new depth and beauty as these details emerge as tiny, significant patterns.

Groups of curious urbanites—in bright clothing and rubber boots—wade into the mountain stream with dip nets, squealing with surprise and delight as wriggly invertebrates emerge from the black ooze.

A young Charles Darwin comes ashore on equatorial islands, midway through a five-year voyage, and carefully observes, then records, the lengths and shapes of the bills of the small birds he finds.

At a predetermined moment, small clusters of biologists begin identifying and counting shorebirds on the expansive mudflats, trying to learn how important this mangrove estuary is to the lives of these intercontinental migrants.

Each of these encounters is an example of the oldest continuous human endeavor—*natural history*, the practice of intentional, focused attentiveness and receptivity to the more-than-human world. Barry Lopez noted that natural history "is as old as the interaction of people with landscape." Simply put, there have never been people without natural history. Every hunting-gathering culture throughout the history of our species practiced careful, deliberate attentiveness to nature—indeed, survival depended on it. Pliny the Elder coined the term *natural history* in the first century AD with the publication of his encyclopedic *Historia Naturalis*—literally, "the story of nature."

Natural history—careful description based on direct observation—provides

Likewise, ecological science today is increasingly cognizant of the social implications of ecological systems. Some fields within ecology, such as conservation biology, are already well along this path. But observational methods are cropping up all over ecological inquiry and also spreading ecology far out into other realms of inquiry. One of Rafe's more unusual projects, for example, is working with an interdisciplinary group the empirical foundation for biology, geology, anthropology, and ecology. The first textbook in ecology, Charles Elton's *Animal Ecology* (1927), began: "Ecology is a new name for a very old subject. It simply means scientific natural history." Most theoretical breakthroughs in ecology have come from thinkers accomplished in field natural history. Witness Charles Darwin and Alfred Russel Wallace, who were both committed naturalists, and E. O. Wilson, who titled his autobiography *Naturalist*. However, academic science in the twentieth century placed abstract theorizing on a pedestal, and devalued the basic descriptive science on which all abstract models are based. The bottom line: without accurate empirical observations, theory is just so much fluff. And, as Harry Greene has pointed out, new natural history information about organisms continually resets research agendas—helping scientists ask better questions and refine theories.

Conservation, too, has always depended directly on natural history. How can we save species from extinction if we don't know where they are, when they're there, and what they're doing? Moreover, for many of us who do field ecology, I suspect, our commitment to conservation has been deepened as much by our direct personal encounters with the world's brilliant wildness as by the data we've collected.

Aldo Leopold frequently deplored the loss of traditional natural history study. In 1938, he delivered an address entitled "Natural History—the Forgotten Science," in which he criticized the new wave of science that increasingly took things apart but failed to explain how they were connected. Leopold objected to the way science forsook natural history when, as he saw it, society needed it most.

Society still needs natural history. Ecology grounded in the best natural history is more dependable, and less vulnerable to political meddling, than science floating on a sea of abstractions. Sustainable resource management depends on natural history insight. And natural history can inoculate society with gratitude for the uplifting beauty of the world, and with the humility this engenders.

of ecologists, anthropologists, psychologists, public health experts, and counterterrorism experts, as well as soldiers, cops, firemen, and spies, to figure out what we can learn from 3.5 billion years of biological evolution for security questions in modern human society (Sagarin 2012; Sagarin et al. 2010; Sagarin 2010). Although some people have called this "Natural Security" project a "new" approach to security questions, it is essentially doing exactly what Allee and Ricketts and many other early ecologists were doing decades ago—taking their observed knowledge about how natural organisms solve environmental problems and connecting it to unsolved societal problems.

But there is also a big difference between what ecologists are doing now and what those long-gone renaissance men and women were doing, and it has to do with the different opportunities available to today's ecologists, arising from new technologies and advances in old technologies that allow us to observe ecological systems in wholly unprecedented ways. Thanks to remote sensing, genomic screening, and animal-borne sensors, to name a few technical marvels, we can now conduct ecology at the very smallest levels of biological organization—at the level of gene-environment interactions—and also at the vary largest levels by observing whole regions of the planet at once. We are even breaking past the boundaries of planet Earth and considering extraterrestrial ecological questions such as, what are the conditions available to support life on Mars?

Rediscovering natural history. Embracing the social sciences. Looking beyond academia for knowledge. Using humans as the focal points of ecological studies and animals as the observers. Adopting technologies once reserved for the CIA and NASA and biotech corporations. All these relatively recent additions to an ecologist's repertoire are collectively stretching and pushing the science into all sorts of new directions. Besides their common roots as essentially observational methods for looking at ecological relationships, is there a way to characterize how these newly acquired tools are affecting ecological science?

## The Domains of Observation-Based Ecology

One way to organize all these different ways of using observations in ecology is to consider the "domain" in which we'd like to conduct science. Steward Pickett and colleagues, who have attempted to define a new twenty-first-century philosophy for achieving ecological understanding (Pickett, Jones, and Kolasa 2007), use the concept of *domain* to mean the "phenomena or scales of interest" of an ecological study. In its simplest form, the domain is defined when we ask, "What is this study about?"

The domain then acts as a filter through which all the data we gather, the theories we consider, and the hypotheses we conceive must pass in order to become part of our study. For example, if our question is "Why are there some purple sea stars and some orange ones?" our domain is basic ecology, and it is likely that theories about environmental justice or data sets on income inequality between coastal human populations will not be all that important in our study.

There are at least four domains in which observational approaches can play an expansive role. First, there is the "purpose" domain which deals with the goals or aspirations of a particular study. Is it basic ecology, aimed at discovering or describing a new phenomenon? Is it applied? Or is it to educate? Observational approaches work well in all of these realms. There are countless basic questions about ecology that can be approached with large amounts of observational data. For example, Rafe and colleagues used 14,000 observations of starfish color and size to reveal that across almost the entire range of the starfish, the ratio of orange- to dark-colored starfish remained virtually unchanged, a completely unexpected and hard-to-explain pattern based on experimentally derived theories of color polymorphism (Raimondi et al. 2007). While experimental approaches also are well suited to addressing basic ecological questions, they have often failed to provide needed insight for applied questions. For example, in the U.S. Pacific Northwest, which hosts Friday Harbor Laboratories, the second-oldest marine biology laboratory on the U.S. West Coast, ecologist Terrie Klinger was frustrated and embarrassed to find that almost none of the ecological studies conducted there over the past century (which were mostly experimental) could help local communities who asked her for scientific advice on conservation and restoration planning (Klinger 2008). Even a simple monitoring scheme to track populations of key species at several sites around Friday Harbor would have been invaluable.

Second, there is a domain dealing with the level of biological complexity that is being studied. Here as well, observational approaches amply fill the space, potentially providing insight at both the molecular level and the global ecosystems level. For example, studies like Brian Helmuth's, which use temperature sensors meant to mimic a living mussel to observe the heat environment of intertidal organisms (Helmuth 1998), can be combined with population-level analysis of where mussels are abundant or not, molecular analysis of heat shock proteins (an indicator of stress in an organism) (Roberts, Hofmann, and Somero 1997; Somero 1995), and even genomic analysis of the genes that regulate stress proteins (Hofmann and Place 2007) to get a realistic characterization of how organisms respond to stress, or to test biogeographic theories such as, "Marine species will be more stressed and show lower populations as their populations are closer to the equator."

Third, there is a domain dealing with the scale of the study in time and space. It is often difficult to conduct experimental manipulations across multiple scales of space and time and impossible to conduct a manipulation of a past ecological state. For example, marine ecologist Bruce Menge and colleagues attempted to replicate Robert Paine's classic "keystone predation" experiments, in which predatory sea stars were excluded from experimental plots on rocky shorelines along the West Coast of the United States, but they encountered an obstacle—the habitat type was much different in California than in Washington, where the original experiments were conducted, confounding some of the results (Menge et al. 2004). Observations fill vast areas of space and long stretches of time, and the variation encountered across those scales is not considered a confounding nuisance but another aspect of the study system to consider. Observations can aid experiments by filling in the dark spaces—those scales where experimental manipulations are unable to shed any light—and they can be used to identify the most important scales in which to conduct experiments.

Fourth, there is an institutional domain that is concerned with the type of people and organizations involved in ecological science. This domain represents a recognition that ecological knowledge is generated not just in academic institutions, but through citizen-science efforts (such as the USA National Phenology Network, described by Jake Weltzin in Box 1.2), through collaborative science projects between managers and resource users, and through a range of new and traditional media. Outside

BOX 1.2

## Citizen Science: Tracking Global Change with Public Participation in Scientific Research

JAKE F. WELTZIN

Within the fields of science and natural resource conservation, unprecedented public access to technology and information (e.g., though online herbariums, species identification tools, mobile applications for image capture and data entry, and community discussions), has enabled people without scientific training to make significant contributions to scientific knowledge. This fact, combined with an increasing awareness by scientists that their numbers are far too few to adequately answer continental- and global-scale questions in a rapidly changing world, has led to the development of "citizen science." Today, in fields as varied as ecology, ornithology, astronomy, and public health, research collaborations between scientists and members of the public are not only helping collect and organize otherwise inaccessible information and data, but are also advancing scientific knowledge that is being applied to issues related to rapidly changing environments across both local and global scales.

One such project that teams citizens with scientists is the USA National Phenology Network (USA-NPN; usanpn.org). The goal of the Network is to establish a national science and monitoring initiative focused on the timing of seasonal biological events—such as flowering, migrations, and breeding. Phenology is a critical aspect of human life—affecting, for example, agriculture, gardening, health, cultural events, and recreation—and of nearly all ecological relationships and processes. Changes in phenology are among the most sensitive and widely observed responses to climate change, and they are relatively easy to observe. People have been tracking phenology for thousands of years for agricultural and cultural purposes, and people still use phenological events, such as the appearance and falling of leaves and the arrival and departure of migratory birds, to track the seasons.

The Network seeks to integrate science and education by encouraging people to make phenology observations that connect them with nature and involve them in the scientific process, and at the same time capture data that scientists are eager to use. Phenology is well suited for this purpose because it is already a primary way that people connect with nature (despite the fact that most people ▸

▶ are unfamiliar with the term *phenology*), and it is an area of rapidly growing scientific interest. The Network has worked with scientists and educators to develop a suite of tools to recruit and retain observers, to share information with and among scientists, educators and managers, and to provide feedback to observers. We are also developing systems for storing, sharing, visualizing, and analyzing the vast amount of data we receive.

Thus, by engaging a willing public in a meaningful scientific activity, in collaboration with expert scientists, the Network confronts the real issue of global climate change and engages the public while providing information critical to sustainability in a rapidly changing world. New phenology networks are also appearing in Australia, Italy, Switzerland, and Turkey and are joining established projects in Austria, the Netherlands, China, and Great Britain, among others. The next challenge will be to integrate, share, and apply data on an international scale to better illuminate patterns and processes that operate across national boundaries.

the academic and (occasionally) the government resource agency parts of this domain, the data gathered and the methods used here are almost exclusively observational, rather than experimental or theoretical. Fisheries managers need simple observations of populations and individual sizes of fish from the fishermen-observers they work with. Communities concerned about local water quality want a way to make observations of the bacterial load or the metal content of the water and share these observations with elected officials or local polluters. And the Discovery Channel wants to show the fangs and blood and terror when a lynx consumes a snowshoe hare before a stark Arctic background, not the trigonometric oscillations of theoretical predator-prey cycles determined by the Lotka Volterra equations. This distinction with academic ecology highlights the fact that the pathway connecting ecological science and social endeavors runs straight through an observational approach to ecology.

Taken together, these approaches give us a lot of ways to study something, almost anything, in an ecological context. It is easy to find examples of all of these ways of using observations in the current ecological literature (see Chapter 2) and in doing so, we are quickly reminded that a diversity of approaches leads to an even larger diversity of findings. For example, just looking at how new molecular observational techniques are used in ecology (see Chapter 4) we find studies that estimate the historic populations of endangered whales (Alter, Rynes, and Palumbi 2007), studies that reveal that wild harvesting such as hunting has effects on both individual organisms and their populations (Allendorf et al. 2008), and studies that examine stress in organisms across their geographic range (Sagarin and Somero 2006).

In some sense, the diversity of findings that comes from observational approaches arises from the methods themselves. Theoretical and experimental studies necessarily test hypotheses under a number of restrictive assumptions and are designed to carefully control variation that doesn't fit within those assumptions. This makes these studies more likely to get results, but the results will accordingly fit into a fairly restricted set. Although observational studies can also be used to test specific hypotheses, they often have an exploratory or speculative component, which means that almost anything can show up in the results. A study by John McGowan and Dean Roemmich that showed a 70 percent decline in zooplankton in southern California, for example, arose from a long-term study designed to figure out why sardine populations were crashing in the mid-twentieth century (Roemmich and McGowan 1995). Another study by Rafe of human poaching impacts on limpet populations was only possible because of a long-term monitoring program to characterize intertidal communities in the case of an oil spill (Sagarin et al. 2007).

As broad and as unpredictable as these possibilities for observation are, it is remarkable that they are still well contained within the science of ecology. Observation-based approaches don't represent a "new" ecology, but rather an evolved ecology that has adapted to external forces and capitalized on abundant opportunities. The change has been fairly rapid. We have seen it occur over just the few years between our graduate studies and our current positions teaching and researching ecology in

Concepción, Chile, and Tucson, Arizona, and even as we write this book we know that the science is continuing to change. Like all evolutionary pathways, the journey of ecology to its current form has a unique history that is inseparable from its present state. And just as with natural organisms, understanding that history, as we attempt to do in the next chapter, is essential to understanding ecology's present form and future potential.

CHAPTER 2

# Observational Approaches in Historical Context

Where did all these observational approaches to ecology come from, and why now, when ecology has had a fairly long run as a respectable discipline using robust theory and controlled experiments, have they begun to emerge everywhere we look? This chapter uses the historical context of how the science of ecology has changed to illustrate that the current changes are both a reflection of an earlier period in ecology and also a unique manifestation, wholly of the current period in environmental history.

## The Roots of Ecology

At its roots, ecology is an observational science, borne out of the work of amateur naturalists and gradually transformed in the late nineteenth and early twentieth centuries into a professional discipline in private laboratories and universities (Fleischner 2005), but it has never been a monolithic enterprise with a single focus and a single pathway for achieving ecological understanding. Throughout this history there has been a tension between broad, expansive views of ecology—represented by efforts to relate observations of natural phenomena to larger questions in biology and sociology—and a desire to make ecology a "rigorous" science, represented by well-controlled tests of theory and predetermined hypotheses.

This tension has driven continual change in the science of ecology, but the changes haven't been random. Like most sciences, and like biological organisms themselves, ecology has grown *recursively*, that is, by building on its own past, and even as it explores new ideas and expands the problem-solving space in which it works, it often returns to previous ideas. In this way the growth of ecology is like the growth of a shelled mollusk—it is a spiral path. The coordinates along this spiral at any given time give ecology its dominant identity, but because it is a recursive form with a traceable history, its past identities are almost always accessible.

Both the dominant themes and continually changing nature of ecology are easy to observe just by clicking through the electronic archives of an esteemed peer-reviewed journal such as *The American Naturalist*. Consider a jaunty paper from 1869 by Samuel Lockwood in volume 3, issue 5, with its vague and innocuous title, "Something about Crabs" (Lockwood 1869). The paper epitomizes both the type of people conducting ecological inquiries in the Gilded Age and the giddy spirit of discovery that drove early ecologists. Using the royal "we" of a proper nineteenth-century gentleman, Lockwood relates some charming anecdotes about various crabs, making literary allusions, drawing wide-ranging metaphors, and often delving into what modern scientists would sneeringly call "anthropomorphism" to describe the crabs as knights in armor, ladies of high stature, or crude strumpets. Describing the spider crab *Libinia canaliculata* (now called *Libinia emarginata*) Lockwood writes:

> She does not covet society, and so withdraws to a cozy grotto, whose walls are green with the tender little fronds of the young sea-lettuce, the Ulva latissima, and the delicately crimped ribbon leaves of the Enteromorpha intestinalis. It did not please us much to see the pert Libinia, with her nippers like little shears, snipping off the velvet lining of the cave. Being indulgent we did not interfere, but left her to her own enjoyment. When we returned, out came Mrs. Libinia in full dress to greet us. On every spine of her uncouth carapace was a green ribbon,—all gracefully waving as she strutted in the open grounds of the establishment. What a sight to look at! And what a lesson in animal psychology! What was the mental process? Was it a device—"a moving grove," like Macduff's, in order to deceive its prey? If so,

what intelligence! Or, was it her vanity? Done just for the looks of the thing! If so, what inexplicable caprice!

Anecdotal reports on the ecology of species such as this were considered crucial to advancing the understanding of the natural world. Considering that these were the first records of many natural phenomena, rather than detailed investigations of minute subfields, authors like Lockwood attempted to make the language relate broadly to familiar ideas.

But as it became more specialized and began generating momentum from within, ecology developed its own internal exclusive language and methods. Fast-forward from the late nineteenth century to the latter part of the twentieth century and in *The American Naturalist* you might find an article like Laurence D. Mueller's "Density-Dependent Population Growth and Natural Selection in Food-Limited Environments: The Drosophila Model" (Mueller 1988). This paper uses a "model system" (as animals with easily manipulated traits such as the fruit fly *Drosophila* and the nematode worm *Caenorhabditis elegans* are called by biologists) to test a highly simplified, mathematically derived theory about how populations of organisms should grow (a "density-dependent" function, meaning that as the density of the population gets larger, the rate of its growth changes, perhaps due to food limitation or crowding). The author argues that even though density dependence has rarely been shown in natural populations, and that "attempts to model density-dependent natural selection in variable environments . . . have yielded odd results," it would nonetheless be valuable to test the formal tenets of density-dependence theory with a highly controlled model system in a controlled environment. Notice the many ways this differs from Lockwood's attempts to connect his direct observations of a relatively unknown organism to things that a literate audience could relate to. Now there is specialized language and there are even specialized organisms. The method is not to compare something that had been observed to an anthropomorphic construct, but to compare something that should have been observed (but hadn't actually been observed yet) to a mathematical construct.

Fast-forward again through the electronic pages of *The American Natu-*

*ralist*, this time to the twenty-first century, and you will find "The Importance of the Natural Sciences to Conservation" by Paul Dayton (Dayton 2003), a marine ecologist well known for his classic experimental studies in Pacific coast tide pools. Dayton here makes a strident plea against reductionism in ecology (and by association even the work upon which he built his early career) and for a renewed focus on teaching observational natural sciences at all levels in order that ecology might restore its ability to be useful for biological conservation. This paper is filled not with theory or experimentation, but with observations—historical photographs of huge sea bass hanging on fishermen's lines, fields of thousands of enormous lobster carcasses, and intertidal rocks carpeted by abalone—images that speak of once-robust natural systems.

Although cherry-picked for illustration, these papers well represent the different phases that ecological science has gone through. Very roughly, these phases include, first, a period of discovery from the late 1800s to the mid-twentieth century, followed next by a half century of increased dominance by theoretical and manipulative approaches that sought to find either general laws in ecology or at least well-supported isolation of causal mechanisms, and most recently the present century of observation-based ecology in which a renewed focus on natural history is the catalyst (if not the complete impetus) transforming ecology.

The early period of formal ecological science was marked by new observations of nature matched to many speculative questions and the formation of basic theories. Early ecologists like Joseph Grinnell asked big, thoughtful questions that required large observational data sets, such as "What is the role of the *accidental*?" (referring to the frequency of, and future prospects for, birds that appeared only one time in a given region's bird list) (Grinnell 1922). Debate began on whether ecological communities were themselves identifiable complex organisms, as Frederic Clements argued (Clements 1936), or merely the result of many individual contributions, as argued by H. A. Gleason in a classic paper, "The Individualistic Concept of the Plant Association" (Gleason 1926). These questions were at the heart of what Sharon Kingsland later referred to as the "sometimes crude but often imaginative and optimistic beginnings" of ecological science (Kingsland 1991).

## Refining Ecology with Experimental Approaches

In the mid- and late twentieth century, ecologists sought greater rigor in their work and hoped to identify consistent laws in their findings. During this period, mathematical ecology flourished in an attempt to simplify and understand ecological complexity using a common language, and more rigorous experimentalism arose to try to move beyond the arguments by association with observed phenomena that appeared to render ecology less "scientific" than other natural science disciplines.

That this phase emerged following the discovery of DNA and the rapid rise of molecular biology seems to be no accident. The very science of biology was redefining itself, and ecology became obscured in the shadow of the bright light shone on molecular discoveries. The tension between ecology and molecular biology was personified in the divergent paths of two young biology professors who both started their careers at Harvard in 1956—the ecologist E. O. Wilson and the co-discoverer of the structure of DNA, James Watson. As Wilson tells it, while the brash Watson was lauded for his zeal in modernizing biology, ecologists such as Wilson were considered nothing more than "stamp collectors" and shunted away into the backwaters of Harvard's biology department even to the point that they were counseled to avoid using the dirty word "ecology" in faculty meetings (Wilson 1994). With so many pressing molecular questions just waiting to be solved, the notion of supporting or hiring more ecologists with their speculative ideas seemed counterproductive to university biology departments.

Perhaps as a result of this ill treatment, and of an earnest desire to demonstrate that ecology could produce the same kind of "results-oriented" progress that marked molecular biology, ecologists (including Wilson) increasingly turned toward manipulative experiments, carried out both in the lab and in the field, which could isolate a limited set of parameters and test theoretical hypotheses. These experiments were largely conducted on small spatial scales and over short periods of time.

In many cases, such as Mueller's investigation of *Drosophila*, experiments were run largely to test theory rather than in direct response to unexplained observations from the field (see Weiner 1995).

The explosion of experimental approaches to ecology provided countless theoretical lenses with which to more clearly appreciate ecological complexity. This period was marked by Wilson and Daniel Simberloff's experiments that used methyl bromide to depopulate entire mangrove islands in order to test the key tenets of island biogeography theory (which sought to explain patterns in the initial colonization and subsequent population trends by animals on remote islands). Robert Paine's decades of work manipulating predator populations along the rocky shores of the Pacific coast became the foundation for ideas about whether ecological communities were controlled by "top-down" forces (i.e., predators' effects on the trophic levels below them) or "bottom-up" forces (i.e., the effect of primary productivity on the trophic levels above), including the development of the "keystone" concept—that certain species had disproportionate effects on the stability of ecological communities as a whole.

These kinds of experiments, and the large body of theory that grew from them, appeared to provide ecology the legitimacy it had lacked in the days when it cowered in the shadow of molecular biology. The often inconclusive musings of Grinnell, Gleason, and Ed Ricketts (which emerged from days-long parties in his Cannery Row laboratory or on long meandering field excursions) were replaced by binary tests that either rejected or failed to reject predetermined hypotheses, and the finality of these conclusions in turn lent ecology renewed confidence. Rather than ineffectually oppose the growing molecular biology fiefdoms, ecologists began to set up their own academic departments. The National Science Foundation also set up its own divisions devoted to ecology, and, though this wasn't formally stated, the focus of funding efforts in these divisions was squarely on experimental ecology, rather than exploratory science aimed at generating new discoveries.

This approach to ecology seems to have brought success. Ecology departments are solid members of all major universities. There are dozens of ecology journals publishing thousands of articles a year. Funding agencies, like the National Science Foundation, have managed to maintain or even increase their millions of dollars of annual funding for ecology despite difficult economic times.

### A New Time of Change in Ecology

These successes notwithstanding, ecology is changing again. Some of this change is not marked formally, but is manifesting itself in ways that don't yet carry weight in the currency of academic progress—hallway conversations and student-run multidisciplinary working groups, student-developed smartphone apps that help people conduct citizen science, and oddball courses like "Holism in Biology," a field course in the Gulf of California jointly taught by neurobiologist William Gilly and literary scholar Susan Shillinglaw.

Even on the principal scoreboard of academic progress—that is, articles in peer-reviewed journals—observational approaches are making a mark. We reviewed over 650 research articles for their methodologies in three high-ranking general ecology journals (based on ISI Journal Citation Reports) across a period spanning our own ecological careers. Excluding reviews and synthesis or opinion pieces, the percentage of articles explicitly using observational approaches to test stated hypotheses (as opposed to merely supporting the creation of experimental manipulations) jumped from 28 percent to 39 percent in *The American Naturalist* and from 38 percent to 55 percent in *Ecology* between 1990 and 2010. In *Ecology Letters*, the most highly ranked journal of original ecological research, the percentage grew from 45 percent in its first year (1998–99) to 54 percent in 2010.

This increase in observational studies can't be explained simply as a result of a declining interest in experimental studies. We found no overall trend in the percentage of manipulative experimental studies, which was similar in *The American Naturalist* (38 percent to 36 percent), dropped in *Ecology* (66 percent to 44 percent), and increased in *Ecology Letters* (33 percent to 45 percent). And in a hopeful sign for an integrative future of ecology, between 6 and 14 percent of articles in the journals studied in 2010 use a combination of experimental and observational approaches, a slight increase in each journal relative to the earlier date of study.

What do these studies look like? The example papers listed in Table 2.1 give us at least a qualitative sense that there is a wide range of scientific advancements, following from basic to applied questions, that are growing out of observational approaches. We have classified a few major types of approaches, such as the use of remotely sensed data or molecular tools, and we found within them a wide range of particular techniques that lead to an even broader range of ecological conclusions than can be robustly drawn from a complex world. Thus even this condensed table, hardly a comprehensive review of all the recent papers that are primarily observational studies, shows that there is much more to observation-based ecology today than just going out into the world and recording what you see.

### Why Ecology Is Changing Now

What factors are driving this shift in ecology, such that the type of studies listed in Table 2.1 are becoming ever more common? One clue to what is happening might lie in the increased interest in restoring the place of natural history—which plays an important role in many observational approaches to ecology—to the life sciences. There are a large number of recent articles in scientific ecology journals identifying insufficient attention to natural history as a major problem for the present and future of ecology, and these complaints are resonating within the science (Greene 2005; Dayton 2003; Dayton and Sala 2001; Weber 1999; Futuyma 1998). For example, Reed Noss's argument against the death of natural history in academic ecology, published in the journal *Conservation Biology* (Noss 1996), resulted in one of the largest outpourings of positive letters in response that the journal had ever received (Fleischner 2005). In 2007 a group of ecologists, educators, and writers formed the Natural History Network to promote the value of natural history within science and in conjunction with societal interests in the arts and community building (www.naturalhistorynetwork.org). Aníbal and Rafe were part of a symposium on the importance of natural history to ecology at the 2009 annual meeting of the Ecological Society of America that drew over 200 audience members and helped to catalyze the formation of a "Natural History" section within the Society.

The focus on natural history captures the essential core of the change in ecological science, but this focus can't explain all of it. For one, esteemed professors' complaints about the loss of natural history in academic journals are, sadly, not likely to drive wholesale shifts in how the science of ecology is conducted. Moreover, these laments are as old as formalized ecological study and have been raised many times through ecology's history. In the late 1800s, for example, future U.S. president Theodore Roosevelt was discouraged as a college student by the dismissal of natural history in academic natural sciences (Millard 2006). In the late 1930s Aldo Leopold used stories of "amateur" naturalists to illustrate the importance of natural history to science (Leopold 1966). The marine ecologist J. R. Lewis penned a letter to the journal *Nature* in 1975 stating pointedly that biologists are unable to help society with the "ecological crisis" because they had not been practicing enough basic natural history (Lewis 1975). Additionally, natural history is not restricted to observational approaches to ecology. Well-known experimentalists like Robert Paine (2010) and Daniel Simberloff (2004) assert that solid natural history is central to creating effective small-scale manipulative studies, even as they downplay the importance of large-scale "macro" ecology and observational pattern finding.

Finally, the changes in ecology we are seeing now—although highly dependent on natural history—are *preceding* the reincorporation of natural history into major curricula, as called for in the many recent critiques. Outside a small number of forward- (or maybe backward-) thinking colleges like Prescott College in Arizona or Antioch University in New England, natural history and organismal curricula have still not been widely reintroduced to biology, and the prospects for field trips grow ever more dim as university budgets shrink and students are increasingly asked to pack more required courses into fewer years of study. (Other colleges with strong natural history curricula are listed at naturalhistorynetwork.org).

We think there are at least four major reasons—both external and internal to ecology—for the reemergence and growing importance of observational approaches.

The first is that the world is changing due to human activity and

**Table 2.1 The Diversity of Observational Approaches in Ecology**

| Observational approach | Example techniques | Example findings |
|---|---|---|
| Measuring basic attributes of organisms, populations, and communities to understand environmental change | Population surveys<br>Population-size structure analysis<br>Long-term monitoring | Populations, when measured at different sites across the geographic range, do not generally conform to the expected distribution (Sagarin, Gaines, and Gaylord 2006).<br>Human exploitation is causing the loss of large, highly reproductive individuals in populations (de Roos, Boukal, and Persson 2006).<br>70 percent loss in zooplankton biomass associated with long-term warming in southern Californian waters (Roemmich and McGowan 1995). |
| Using ancient records to document environmental change | Dendrochronology—the study of tree-ring patterns<br>Reconstruction of fossil records<br>Sampling biological sediments (pollen and shell deposits)<br>Sampling shell middens from human settlements | K-T extinction was likely driven by extraterrestrial impact (Alvarez and Asaro 1990).<br>Trend toward more specialized (hypercarnivory) diets leads to increased risk of extinction in mammals (Van Valkenburgh, Wang, and Damuth 2004).<br>Lake charcoal sediments demonstrate that natural age range of boreal forests is not found under current management practices (Cyr et al. 2009). |
| Using historical records to document environmental change | Examination of matched photographs (landscape, vegetation changes)<br>Georeferenced maps<br>Examining naturalists' notebooks<br>Phenological records<br>Natural history museum collections | Fire suppression and climate change have dramatically altered landscapes (Webb, Boyer, and Turner 2010).<br>Wolf extinction in Yellowstone produced the decline of cottonwood populations (Beschta 2005).<br>Historic records suggest large loss of Caribbean sea turtle nesting beaches (McClenachan, Jackson, and Newman 2006). |

| | | |
|---|---|---|
| | Nontraditional amateur records (sailing records, restaurant menus, contest records, artwork, literature, etc.) | Genetic material from museum specimens can be used to track origins of pesticide resistance (Hartley et al. 2006). |
| Using remote observations to document patterns and change at multiple scales | Satellite images (multispectral and hyperspectral resolution) (Turner et al. 2003)<br>LIDAR (*l*ight *d*etection *a*nd *r*anging)<br>Low-elevation aerial photography<br>Camera traps to track elusive animals | Remote sensors can effectively detect invasive species across different spatial scales (Underwood, Ustin, and Ramirez 2007).<br>Humans have modified land use and landscape configuration exponentially in recent decades (Tang, Wang, and Yao 2006). |
| Observing how organisms and natural objects experience ecological phenomena | Animal-based data-logging tags<br>"Critter cams"<br>Physical-factor data loggers | Satellite tags reveal that Atlantic bluefin tuna cross stock management boundaries, providing insight for better management (Block et al. 2005).<br>Animal-based cameras reveal previously unknown behaviors in sea turtles and penguins (Moll et al. 2007).<br>Heat stress experienced by intertidal organisms does not follow the expected latitudinal gradient (Helmuth, Kingsolver, and Carrington 2005). |
| Using molecular tools to understand relationships and dynamics of populations | *In situ* assays of RNA:DNA ratios and heat shock protein (HSP) concentrations<br>Genetic population structure<br>Phylogeography<br>Genomics | Historic gray whale populations were likely to have been much larger than expected from fishery records (Alter, Rynes, and Palumbi 2007).<br>Exploited species show long-term genetic effects on individuals and populations (Allendorf et al. 2008).<br>Pyrosequencing of eukaryote DNA in estuarine sediments used to characterize diversity and differentiate contaminated and uncontaminated sites (Chariton et al. 2010). |

(continued)

**Table 2.1** *(continued)*

| Observational approach | Example techniques | Example findings |
|---|---|---|
| Using natural experiments to test ecological hypotheses | "Controlling" variables through large-scale comparisons in space or time, before/after catastrophic event, in/out of protected reserve<br>Comparative observations along a gradient | Roads facilitate the movement of invasive plants into protected areas (through increased disturbance and propagule pressure) (Pauchard and Alaback 2004).<br>Historical, political, and economic comparisons between Haiti and the Dominican Republic reveal causes of relatively greater ecological destruction in Haiti (Diamond and Robinson 2010). |
| Using large data sets collected for multiple purposes to draw generalizations | Multidisciplinary assessment of environmental impacts<br>Meta-analysis of published studies | Global review of ocean impacts shows no part of the ocean is unaffected by humans and 41 percent is strongly impacted (Halpern et al. 2008).<br>Multiple studies of population and community change reveal a consistent, globally distributed signal of climate change (Parmesan and Yohe 2003). |
| Learning from people who interact closely with the environment | Assessing indigenous people's knowledge, legends, and myths<br>Assessing farmers' and fishermen's knowledge<br>Taking oral histories<br>Social science surveys | Recognizing the existence of "cultural keystone species" may improve conservation efforts (Garibaldi and Turner 2004).<br>Differences in knowledge between generations of artesian fishermen can be used to infer historic population trends (Saenz-Arroyo et al. 2005). |

these changes are affecting natural systems at larger spatial scales and across longer temporal scales than ever before in human history. As Peter Vitousek and colleagues (Vitousek et al. 1997) have pointed out, no landscape is without significant human alteration, a conclusion later extended to Earth's ocean environments by Ben Halpern and colleagues (Halpern et al. 2008). Even being able to reach these startling conclusions required the compilation of large numbers of observational studies, but the more fundamental problem is how few experimental studies of the last several decades (which were often conducted in artificial microcosms or on "pristine" scientific nature reserves) explicitly included the human behaviors leading to environmental degradation as a variable of interest. As a thought exercise, we can imagine a set of well-controlled experiments that can determine the varying role of a parasite on a voracious insect that favors a particular rare plant species, yet we can also imagine that if humans (through a complex set of social, economic and political machinations) decide to put a Walmart down on one of the last remaining patches of the host plant, humans, and not the parasite, become by far the most important player in the ecology of the plant. For better or worse, in most cases human behaviors cannot be experimentally controlled or manipulated. Observational studies, which in this example might include historical analyses of the plant's former and current distribution or social-ecological assessments of the motivations and alternatives for placing a Walmart on top of a rare patch of plants, are needed in conjunction with experimental studies of small-scale functional studies in order to understand the actual, as opposed to the idealized, ecology of the system.

Second, in no small part because of these massive environmental changes, the attitudes of ecologists, especially new students, have changed. Today's students are determined to help solve environmental challenges and they are almost effortlessly interdisciplinary. Their attitudes in many cases are trickling up to their professors, who are also increasingly open to applied and interdisciplinary studies. For example, Rafe became interested in social science methods of observation, such as interviews with resource users and public opinion surveys, chiefly because that's what his students were doing to understand things like whether eco-labeling programs would help fishermen and the environment (Goyert, Sagarin, and Annala 2010). The more applied questions (and the larger scales at which they apply) that these students and their professors are asking require relatively new observational tools borrowed from other fields, such as structured human-subject interviews and linguistic analysis of policy documents, as well as intensified use of observational tools, such as GIS and spatial statistics, that are already widely used in ecology.

Third, an unprecedented opportunity to study ecological systems across scales of space and time has become available through new technologies, the accumulation of long-term data, and the simple passage of time relative to historical descriptions of ecological systems. Questions about ecological systems and changes to those systems that never could have been answered previously, can now be unraveled—just by observation. New technologies create opportunities to answer straightforward but previously unanswerable questions like "Does anything live in the deepest reaches of the sea?" and "Where do bluefin tuna go?" Even questions that hadn't previously been asked are being answered. Genomic studies, such as screening for microbial genomes within a drop of seawater, have revealed whole new realms of biotic diversity (Breitbart et al. 2007). There are also opportunities to answer riddles that would have been impossible for even the most astute ecologists from earlier times to solve. For example, both Henry David Thoreau and Aldo Leopold took extensive observations of phenological (timing) phases of plants and animals in their environment, but neither lived long enough or had the understanding of global change to know that these records would one day become essential indicators of climate change's effects on natural systems (Nijhuis 2007; Bradley et al. 1999).

Fourth, new observational approaches to ecology in reference to problems as diverse as climate change, ocean acidification, invasive species, and endangered species management have already amassed a proven record of success. Observational studies have been critical when we want to understand how large-scale, high-impact, and non-replicable events affect ecological systems. For example, of the 143 non-redundant papers cited as evidence in the two seminal reviews of how natural systems have already responded to climate change, 132 (92 percent) were observational studies (Sagarin and Pauchard 2010). Likewise, the ecological effects of the Chernobyl nuclear disaster (Bradbury 2007), the protective services provided by intact mangroves during tropical storms (Granek and Ruttenberg 2007), the fate of the oil from the Deepwater Horizon blowout in the Gulf of Mexico (Camilli et al. 2010), and even the effects of airplane travel on climate (as determined, cleverly, by studying comparative weather patterns in the few days after the 9/11 terrorist attacks when no commercial airlines were flying over North America [Travis, Carleton, and Lauritsen 2002]) have all relied on observational studies. But the extremely small scale is also being captured by the new observational ecology.

It isn't just coincidence that these factors came together to change ecology. Rather, they are an interacting set of attractors that generate self-sustaining momentum and build recursively off one another. That is, as observational technology and the passage of time affords us greater opportunity to see large-scale change in the world, ecologists and those inclined toward the study of natural systems (i.e., those with a strong sense of what E. O. Wilson calls *biophilia*) develop further motivation to understand and do something about those changes, which requires them to acquire knowledge or develop relationships with experts in a broad range of fields, and, in many cases, to make the most out of observational data sets. As the results of their work gain acceptance and are exchanged into the currency of academia—such as publications in *Nature* and *Science*, or the awarding of MacArthur "genius" grants to scientists like Barbara Block, who developed remote tags to track bluefin tuna (Block 2005), and David Montgomery, a geomorphologist who shed the boundaries of his field to understand the complex relationships among people, the environment, and salmon (Montgomery 2003)—there is not only a built-in incentive to continue such lines of exploration, but to expand the opportunities (develop new long-term monitoring programs, dig up more historical data, launch new satellite platforms, deploy more "critter cams"). These opportunities then provide a clearer picture of environmental change, and also draw new people—students, citizen scientists, and long-term observers of the natural world like fishermen and farmers—into the expanding realm of ecological inquiry. In other words, like the longer-term history of ecology, the current trend in observation-based approaches is following a spiral path, growing recursively and expanding its influence into other fields.

The next section of this book is about how to get wrapped up into this spiral of change that is occurring in ecology. It starts at the core of our own human senses, which are both the first, and the most underutilized, tools for achieving ecological understanding. After talking up the power of our own senses we then concede their limitations and suggest that we can expand our understanding of ecological complexity by entering into symbiotic partnerships with observational technology and computers, providing that we take care that such partnerships don't eclipse our innate observational abilities. Finally, we suggest that these innate and acquired observational tools are by no means limited to scientific ecologists. If ecology is going to open itself to all sorts of new observational methodologies, it must also open itself to the fact that the best ecologists might not know they are ecologists at all.

PART II

# USING OBSERVATIONS IN ECOLOGY

The core activity of ecology is "observation." Humans have been keen observers of the natural world as long as we have lived on earth. Obviously, this has served a critical purpose—observing immediate dangers as well as the relationships of biological organisms with one another and their changes between and across seasons were all essential to survival. But from the first recordings of human observations—pictographs, carved fetishes, and cave paintings—we can see also a great awe and wonder at the natural world. We don't think these two drivers of human observation—the informational content and the emotional returns—need to be separated when considering an observational approach to ecology. Nowadays, more than ever, observations of our changing planet are profoundly important to our survival, but the love of nature itself and our inherent curiosity about it are essential motivations for the best observations.

We start in Chapter 3 by illustrating the importance of utilizing multiple observational senses to achieve ecological understanding. Using examples from nature and from different remarkable observers of nature, we show that abundant ecological information exists beyond our visual field, and that many ecological secrets are only revealed when we stretch out our senses and continually practice the skill of observing. In Chapter 4 we expand our observational capabilities even further by reviewing the wide array of new technologies that allow us to expand our innate observational senses into previously unfathomable expanses of space, time, and sensory spectra. Remote sensing allows us to view phenological changes and waves of species invasions across entire regions and at different spatial scales. Molecular biology, which in the twentieth century caused a deep rift between naturalists and supposedly more "rigorous" biologists, lends itself to observational approaches that are now being fully integrated with ecological studies. Animal-borne sensors are essentially turning animals into observers of the natural world and in the process are immediately overturning long-standing assumptions about the basic ecology of even well-studied organisms. In Chapter 5 we argue that the resurgence of observational approaches presents an unprecedented opportunity for opening ecology to a wider population. This more inclusive ecology is becoming evident in all aspects of the science, including much greater deference to local and traditional forms of ecological knowledge, the emergence of citizen-science programs as both an educational tool and a rich data source, and, critically, an embracing of social science methodologies. Here we outline the sometimes surprising range of data sources that have already contributed to our modern understanding of ecological change.

CHAPTER 3

# Using All the Senses in Ecology

This chapter is about how our sensory abilities to perceive nature are essential to an observational approach to ecology. These sensory abilities are both universal to humankind and at the same time unique to different individual humans, based on their personal history, abilities, and motivations. Our senses are our most elemental tools in building an observational understanding of ecological relationships, but they are often underutilized and sometimes viewed with skepticism in a scientific context. In this chapter we show how each of the senses can contribute to scientific ecology. We use the experiences of past and present ecologists to argue that the personal nature of how we utilize our senses can be an asset that motivates us to explore the natural world and opens us to new ecological discoveries.

## Sensing Nature

The human capacity to observe the natural world is highly diverse and it is variable through time, and therefore it can be heightened or dampened, and it can be improved through experience. Geerat Vermeij, a paleobiologist, has written that "the skill of observing—and it is a skill, to be honed and perfected—must be taught and encouraged. It is something that every science student must possess" (Vermeij 2002). Vermeij here puts observation at the core of scientific literacy, and this sentiment has

BOX 3.1

## The Importance of Sensation

GEERAT J. VERMEIJ

Scientists revel in a way of knowing that uncovers an approximation of verifiable truth through observation, evaluation, and inference. We have refined this method, but scientists did not invent it. From the beginning, living things have sensed, interpreted, and responded to circumstances that could make the difference between life and death, success and failure. Informed by their senses, organisms embody a hypothesis of their environment; and when this hypothesis is tested—when an organism's structure, physiology, and behavior work adequately—it can be improved as the body and the environment as sensed and interpreted by the organism feed back on each other, both through immediate effects and over evolutionary time through natural selection. A profound parallel exists between adaptive evolution and the more purposeful scientific way of knowing. Environment and hypothesis converse, whether in the body of an adapted organism or in the mind of a human being.

This parallel highlights the essential, and increasingly ignored, role of sensation—of observation with the brain in gear—in learning about the world. There is nothing like being puzzled by a chance observation to awaken curiosity, nothing like carefully listening and looking and feeling and smelling to conceive ideas,

been echoed by other prominent ecologists who lament the loss of natural history–based classes at all levels of education (Dayton and Sala 2001). His focus on observation is not surprising, as he works in a field where few experiments are possible and large observational data sets gathered from the fossil record must be brought to bear. Yet this focus on observation has certainly not handicapped Vermeij as a scientist. He has not only contributed substantially to ecology and the study of evolution, but also to a wide range of social studies—from economics to security—in the course of his career (Vermeij 2004).

What is surprising is that he has been blind from a young age, dependent upon his tactile sense to "observe" the natural world, as he describes in Box 3.1. Many sighted scientists have marveled at how Vermeij's tactile ask questions, and formulate theories of how the world of things and people works.

As a blind boy, I daydreamed about the tropics. There were fine descriptive books about the lush vegetation, colorful birds, and beaches strewn with wonderful shells; and I have an inkling that some of the world's great natural scientists—von Humboldt, Darwin, and Wallace among them—were so stimulated by the things they saw in those equatorial regions that they changed our very conception of the world of living things. But it took first-hand experience—literally, of course, and first-ear and first-nose experience, too—to make me ask questions and ultimately perhaps to understand the wet, warm forests and the thrillingly diverse reefs and sandflats of the tropics. It was these experiences, informed by the senses, that helped shape my scientific worldview.

Children today are taught to take tests. They study virtual representations, and are in the position of passive consumers of films and real-time feeds as others explore the world through their own sensibilities. The unfamiliar, insofar as it is accessible at all, comes packaged and manufactured. Is it any wonder that our curiosity withers and our contact with the world atrophies?

Good observation is a skill, to be honed and nourished and improved. It is like reading or writing: the more you do, the better you get at it, and the more the world opens up to you. Educators, pay attention.

observations channel intricate details and life histories from fossilized shells that they themselves could not see with their own eyes. His story is jolting to the sighted among us because most of us immediately think of our visual sense when we think of "observation."

Vermeij's story begins to reveal the astonishing adaptable capacity of our observational abilities, and this capacity comes into still greater clarity when we consider another highly accomplished teacher who also happens to be blind. Daniel Kish observes the world primarily through sound. He taught himself at a young age to use his own sonar system, by making audible clicks and listening for their echoes (Kish 2009). Soon he could identify the shape and materials of objects in his surroundings. He could get himself to school and through his day on his own. He could even ride a bicycle. And now he teaches other blind students to do the same, primarily by observing the world with their ears.

These two remarkable observers of the world highlight the wide range of unexplored possibility for most of our observational senses and the limitations of relying too much on visual observation. It should be self-evident that ecological science requires all of our senses. After all, the ecological world is a swirling miasma of sensory information. The relationships among plants and animals and microbes and the physical world around them are all crafted in touch—deadly bites and hypersensitive whiskers; in sights—flashes of warning colors and cloaking camouflage; in smells—odors that attract and repel; in sounds—alarm calls and mating songs; and in tastes—the bitterness of protective chemical compounds and the sweetness of nectar. Likewise, ecological study is punctuated with remarkable uses of the senses, visual and otherwise.

The full use of our senses can make the difference between simply getting through the immediate challenges of our daily lives and developing a deep and scientific ecological understanding. Going out on a hike with really good birders is an *ear*-opening experience. Most of the birds they identify they don't even see, or get only a fleeting glimpse. Instead, they have catalogued an immense collection of calls, songs, and variations so that they can identify by sound not only the species of the bird, but also its gender, its behaviors, and even how much it reveals about the presence of other animals in its vicinity.

Indeed, sound creates its own ecology, affecting relationships between individuals of the same species and other species, but because we downplay sound observations, we are often unaware of this sound ecology. David Dunn, who has spent his life cataloging ecological "soundscapes" by using a variety of cheaply built homemade sensors, has discovered through sound whole new relationships that were overlooked by even the most experienced biological experts. His work on bark beetles, which have damaged millions of acres of forests in the United States' southwest has revealed that they are vulnerable to auditory stimulus (www.fs.fed.us/r3/resources/health/beetle/). When noises in the right frequencies are played near them, male bark beetles cease to consume the inner bark of trees and instead attack females, suggesting a potential pathway for bark beetle control.*

Sometimes nature must be touched to be understood. When Rafe takes children on tide-pool tours he always ensures that they touch as much as they can. The parents, teachers, and Girl Scout "Den Mothers" who chaperone them are often taken aback. Usually the kids have been given strict instructions beforehand not to touch anything. These prohibitions undoubtedly come from a well-intentioned desire to protect the environment, but they are unnecessary and counterproductive. They are unnecessary because it's actually rather difficult for an individual human, even in a group, to do much permanent damage to most ecosystems just by touching. There are exceptions, of course—stepping on live corals while diving, hiking across delicate cryptogam soils in the desert—but many organisms, especially those that have evolved to withstand some of the harshest forces and stressors on the planet, are quite robust. At the same time, prohibitions against touching are counterproductive because you can't learn anything from nature without touching it. Touching is a natural way for children to explore their world and it instantly creates all sorts of ecological questions. Why are those tide-pool plants stiff and these floppy? Why is this green anemone covered in shells and debris and that other one totally smooth?

An amusing note from Ed Ricketts's journey to the outer shores of the Pacific Northwest indicates that even well into his career as a biologist he used his sense of touch to make new discoveries: "Another big octopus. I've often wondered if octopi ever bite. Today I found out. Yes, they do, they certainly do" (Ricketts 2006).

Taste is a trickier proposition when it comes to ecological study. Natural organisms contain a wide range of defensive chemicals that in some cases can be toxic to humans, so it is probably not a great idea to

*"Here Comes the Sound," interview on *Living on Earth* radio program, aired 10 February 2010 (loe.org/shows/segments.htm?programID=10-P13-00009&segmentID=6, accessed 14 October 2010). This interview with David Dunn and researchers Richard Hofstetter and Reagan McGuire reveals both the power of interdisciplinary collaborations and the enormous gaps in our understanding of sensory ecology.

encourage Girl Scouts to taste the animals they find in the tide pools. And because we cannot taste with our arms like an octopus, using our sensitive mouth parts for ecological purposes can be a dangerous proposition, especially if one happens to be doing ecology in harsh environments—deserts and intertidal zones, for example—where organisms are well armored. Nonetheless, there are opportunities for bringing even our sense of taste into ecology. Among the hundreds of algal species that inhabit the shores of the Pacific coast there are some, like those in the genus *Osmundea* (formerly *Laurencia*), that can be identified by their distinctive taste. One of the great algal scientists of the Pacific coast, Isabella Abbott, took mischievous pleasure in officially naming one distasteful member of this genus, *Laurencia blinksii,* after Lawrence Blinks, the rather sour former director of Hopkins Marine Station in Pacific Grove, California. Many plants have distinct tastes that can help us identify them, and botanists and phycologists who have taken the time to learn the difference between toxic and merely distasteful plants have long used taste to identify plants and algae.

Olfactory senses drive numerous ecological interactions. Many mammals use scent marking to claim territory or communicate information about dominance hierarchies. Salmon and other anadromous fish use olfactory cues in part to guide them back to their natal stream reaches (Montgomery 2003). Lobsters use scents in complex mating rituals (Corson 2004). Indeed, smells play a lesser, but not insignificant, role in ecological observation, and for terrestrial plant ecologists smelling can be a good surrogate for tasting. Good animal trackers report that they can smell the musk on a tree that has been rubbed by a buck. The more humid areas in a forest can be recognized just by the smell. Even very young children can be taught to differentiate plants by their smells. In dry ecosystems such as the Chilean matorral (Mediterranean shrublands), plants have developed particularly strong biochemical compounds that incidentally allow for easy identification of the species (Muñoz, Montes, and Wilkomirsky 1999; Hoffmann 1989). In coastal ecosystems, the same distinctive *Osmundea* algae that can be identified by taste can also be located in a dark nighttime tide-pool excursion by their acrid odor, and a common intertidal sponge,

*Halichondria panicea*, has a very distinctive sulfurous odor that can be used to confirm its identity.*

## Training Our Senses for Ecology

But how is using our observation abilities as ecologists any different from simply going through our everyday world seeing and hearing and touching and smelling and tasting things? Here again, Daniel Kish, the blind teacher who uses echolocation, provides some insight. When an interviewer once asked him if sighted people could learn echolocation, Kish expressed his doubts, noting that heavy doses of "motivation, necessity, and practice" were required, none of which are very prevalent in sighted people with respect to echolocation (Mithra 2004). Fortunately, ecologists excited about their subject matter do tend to get highly *motivated* and do *practice* their observational skills a lot. One of the most useful means of practice for an ecologist is to keep a field journal, as ecologists Anne Salomon and Kirsten Rowell have shared with us and described in Box 3.2. And *necessity* comes from the nature of the work itself—if you are not a good observer, it's nearly impossible to find anything interesting at all, let alone to draw ecological insights or unravel environmental mysteries. It is this trio of ingredients that separates truly useful ecological observations—the ones that might help us achieve a scientific understanding of ecological complexity—from the standard observations that get you through the day.

The most obvious evidence of this appears when we take different students out into the field. There are certain students who inevitably see far more than the others. They are the ones who spot the elusive horn shark on a snorkeling trip through the kelp forest or are able to identify tree species in the forest canopy. We find it remarkable that these sharp-eyed students often have never been to the site we're exploring or even the same type of ecosystem. The students with "good eyes" invariably share certain characteristics. They are always the students who spent

*The examples of distinctive smells and tastes of intertidal organisms, as well as the story of "Izzie" Abbott's naming of the algae, come from two great West Coast naturalists, John and Vicki Pearse.

BOX 3.2

## The Art of Ecology: How Field Notes and Sketches Offer Insights into Nature

ANNE SALOMON AND KIRSTEN ROWELL

For millennia, humans have used drawings to communicate and document nature's mysteries. From the Paleolithic pictographs of Pech-Merle (Pruvost et al. 2011) to the meticulous notebooks of twentieth-century biologists like Joseph Grinnell, the tradition of field illustration and note-taking of the natural world has proven a worthy means of developing ecological insights. Whether on rock or in spiral-bound notebooks, field sketches and notes are an invaluable initial step toward illuminating the emergent queries and synthetic understanding that lead to predictive science. In ancient times and still today, art and science have been used in a dialectic to seek those fundamental ecological truths upon which our survival depends.

Our cognitive processes are sharpened by the practice of distilling and recording *in situ* observations of the natural world. By detailing the intimate aspects of a species' morphology (Fig 3.1), or the nuances of an ecological process (Fig 3.2), field drawings and notes can help refine our thoughts, crystallize our inklings, and enrich our ecological intuitions even before we have a clear understanding, let alone the words, to describe the phenomena we are witnessing. Simply put, drawing can improve our seeing (Edwards 1999). By archiving observations, humans etch in their minds ecological ideas in their infancy. And yet, just as notes and drawings can engender ecological intuition, so can they help decode it.

The practice of identifying and sketching key features of a species, landscape, or ecosystem propels the simplification of a complex nature down to its essential parameters from which accurate generalizations about nature can be made. This after all, is the main objective of ecology. It is these raw recordings of nature's indispensable elements that can provide deep insights leading to novel and testable hypotheses. Skillful observations are the foundational building blocks of the research process, and thus their careful documentation is a fundamental aspect of ecological literacy.

The benefits of archiving the context and characters of our research sometimes manifest down the road. Much like museum curators, ecologists who collect, organize, and preserve observations and ecological insights have a timeless resource for future study. Field notes and illustrations, regardless of how rudimentary, can capture fleeting impressions that in later years offer a basis for reflection and comparison (Greene 2011). Georg Steller's sketches of the

**Figure 3.1**

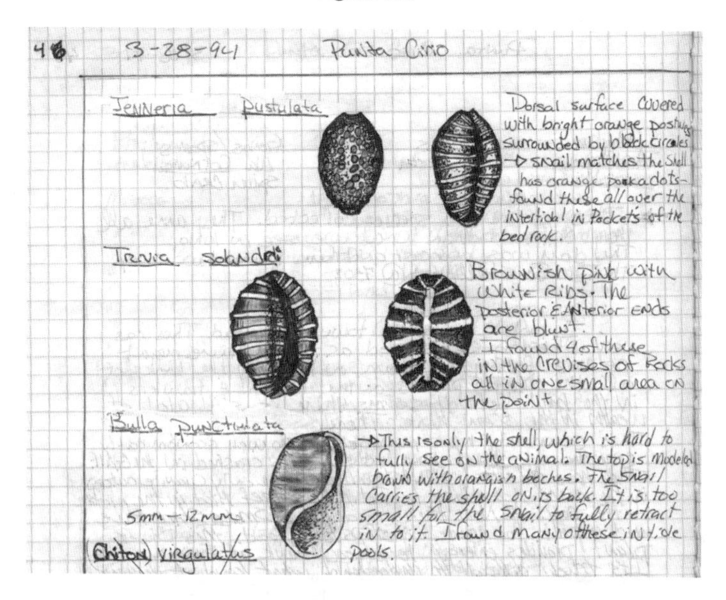


*Undergraduate studies on shell morphology and ecology of intertidal mollusks in the Gulf of California. Drawn by Kirsten in 1994.*

formerly abundant sea cow offer an example of how archived field drawings can reveal ghosts of ecosystems past (Dayton et al. 1998) and avert sliding baselines (Pauly 1995; Jackson et al. 2001). Similarly, petroglyphs and pictographs that capture natural history observations made thousands of years ago offer ecological baselines in a universal language that bridges cultures and transcends time.

Today, however, with our pressing need to make ecology a more predictive science, the art of field illustration and note-taking has taken a back seat to modern ecological modeling and molecular techniques (Dayton and Sala 2001). This cultural shift has changed the skills we value and thus the skills we teach in ecology such that field-sketching and note-taking are in jeopardy of becoming expatriated. Although scientists now have a greater capacity to model ecological systems than ever before, we often lack the basic natural history information needed to parameterize increasingly complex models (Tewksbury et al. in review).

Yet natural history is the starting point for all progress in ecology. Without it, even the most complex models can yield inaccurate predictions that, when applied to solve real-world problems, can have unintended consequences. As we advance ecological technologies we should simultaneously continue and ▸

**Figure 3.2**

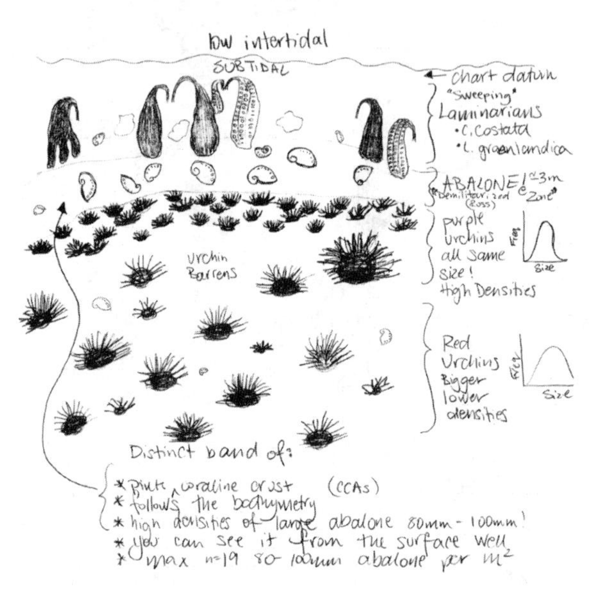


*An as-yet undescribed phenomenon in the shallow subtidal rocky reefs of Haida Gwaii, British Columbia, Canada, where a feeding front of herbivorous purple and red sea urchins meets sweeping Laminarian kelps at a specific depth that likely shifts with the tides, creating a distinct pink horizontal band of smooth crustose coraline algae with unusually high densities of exposed, openly grazing northern abalone. Drawn by Anne in 2010.*

▸ advance the simple art of field illustration and note-taking. While satellite imagery, hierarchical modeling, and stable isotopic signatures have become vital tools in the contemporary ecologist's tool kit, so should the skills of field-sketching and observation. If, as Ramon Margalef (1997) once said, a naturalist is more a poet than an engineer (Margalef 1997), then the skills of an artist are equally worth training.

abundant hours in the outdoors as children. Many of the good observers have been raised in rural places or have been out in the field since childhood. They may have grown up creek-walking in forests, or fishing with their parents, or just finding solitude in small suburban pockets of woods, but they'll quickly find themselves right at home when taken to a forest, the desert, or the coast. These "gifted" students are also almost always the first ones into the field and the last ones out. They will be up to their thighs in marsh muck before half of the students have even gotten out of the van. On a cold night snorkel in Pacific waters, they will outlast the other students and their shivering instructor by 20 or 30 minutes.

In an article he wrote before he became famous for books like *The Tipping Point* and *Blink*, Malcolm Gladwell noticed a similar phenomenon among "physical geniuses"—great musicians like Yo-Yo Ma and great athletes like the hockey player Wayne Gretzky (Gladwell 1999). What Gladwell found was that these people put in an enormous amount of time practicing their craft, even when they were already far and away the best in their fields. All that practice gave them a huge storehouse of "chunked" memories (Churchland 2004) that could be recalled and used to map out even novel situations. (Gladwell expanded this idea later in his book *Outliers*, where he cited research suggesting that, in addition to fortunate circumstances and a minimal basal intelligence, the exceptional practitioners among us possess an enormous history of practicing their craft—typically at least 10,000 hours [Gladwell 2008].)

In this light, ecological observing seems not unlike practicing scales on the cello or slap shots on a hockey rink, or learning to echolocate or see by touching. It is a necessary skill, though not sufficient in itself, that creates the opportunity for mastery of complex systems. In forestry, a colleague of ours has coined the term "forest hours" (Eduardo Peña, pers. comm. to AP), in the same way airplane pilots keep track of "flight hours," to explain how just the continuous exposure to forest ecosystems may help a professional to acquire a broader and deeper understanding of forest dynamics.

So what are the neurological and psychological bases for such learning process?

Learning through sensory observation includes at least three key components: inherent skills (our genetic background), previous experience and knowledge (especially childhood experiences), and cultural background (how much was free-ranging observation permitted and encouraged in the course of our development?). As with having an ear for music, some of us are better or worse at observing phenomena. However, early stages in our development are crucial to refine our senses. Undoubtedly, a child living indoors for the first three years of his life will have a much harder time recognizing bird songs. Previous experience and knowledge will not only develop our skills but will also help us create a cognitive network or mental map by which new information can be integrated into the complex puzzle of all our previous observations (Bransford 2000). Finally, our cultural background and values gives us the motivation to pursue the observation of nature. Families where nature is honored as an important part of life will foster a much stronger motivation in their children to be aware and alert to nature's signals. A common theme to both Geerat Vermeij and Daniel Kish's childhood experiences as blind observers was that their parents wholeheartedly encouraged them to go out and explore the world.

At the more personal level, and again as illustrated by great observers like Vermeij and Kish, we each use senses in a way that reflects our own history and proclivities. The uniquely personal abilities and motivations of individual ecological observers have been powerful determinants in the progress that ecology has made as a whole. Too often, though, we assume that "personal" means "biased"—a word that carries negative connotations in science. In one sense, we are all biased. We chose to study forests or coastlines or deep-sea hydrothermal vents, and these decisions clearly color our particular viewpoint. Although Geerat Vermeij is careful to avoid portraying his blindness as a handicap, there is no doubt that the special way he uses his senses shaped the type of scientist he is today and helped define the range of discoveries he has made.

But "bias" in the dangerous sense of leading to false or misleading results is ultimately a function of the study design and execution, not of the investigator himself. Both of us have experienced the wide range of observational skills among our students and this can be disconcerting if multiple observers are gathering data for a large project. One thing we can do is devise tests both to train our observational skills and to ensure a certain level of consistency in our observations. For example, students or new field technicians can be trained by having them estimate the percentage of canopy cover on simulated plots with a known percentage of cover values before they take data from the field; this training can help everyone understand a common language of observation, but it still won't remove all bias. A long-term-impact study of forest health conducted by the University of Arizona has a "training plot" among several dozen large observational plots in the field site where new members of the research team learn the field techniques. One way to look for bias in this kind of study is to analyze the field data against the training plot to see if there are consistent differences (such as higher or lower scores for canopy cover) in the plot scored by inexperienced researchers. Pretending that bias doesn't occur, or that we've washed it all away by, say, randomizing our observations, won't get rid of it. Rather, dealing with bias requires another keen observational ability—that is, the ability to conceive of and find potential sources of bias within a study or a data set.

Sometimes we just need a test to overcome the skepticism that one can attain remarkable observational skills just through the day-in, day-out process of field work. Christopher Norment, a bird ecologist, wrote in his memoir *Return to Warden's Grove* of both his confidence and skepticism regarding his own "intuition" about where to find elusive sparrow's nests (Norment 2008). While he had to rely faithfully on this intuition to get anything useful out of his field seasons in remote northern forests, when he returned home to the Midwest he conducted an unusual, but scientifically rigorous test of his skills. He had his graduate students take him blindfolded to an unfamiliar field and time his nest findings relative to uninitiated observers, and he found that his intuition was indeed correct—he had become a great nest finder.

Ed Ricketts argued that the intensive, personal process of observation was both an antidote to many forms of bias and also a driving force behind some of the most robust ecological theory. He advocated a process of observing first and then building multiple observations into holistic theories—what he called "non-teleological" thinking—rather than the deductive approach of starting with a theory and then dividing it up into smaller parts. It was through this approach of starting by asking "what" a system was, rather than "why" a system functioned as it did, that the biased assumptions all investigators bring could be avoided (Ricketts 2006).

But it was—according to Ricketts and his good friend, the author John Steinbeck—the combination of this observational approach to science and the deeply personal, emotive aspect of natural science that truly led to scientific progress. They seized on a passage from Darwin's *Voyage of the Beagle* in which Darwin exclaimed of Valparaiso, "When morning came, everything appeared delightful. After Tierra del Fuego, the climate felt quite delicious—the atmosphere so dry, and the heavens so clear and blue with the sun shining brightly, that all nature seemed sparkling with life" (Darwin 2004). Ricketts and Steinbeck speculated that this bidirectional nature of natural history was at work:

> Darwin was not saying how it was with Valparaiso, but rather how it was with him. Being a naturalist, he said, "all nature was sparkling with life," but actually it was he who was sparkling. (Steinbeck and Ricketts 1941)

While there is a romantic quality to this interpretation, we should remember that Darwin was sick as a dog during nearly all his time at sea, and though he made many contributions to marine biology, they came largely through the painstaking hours he spent in his country home teasing apart the taxonomy of barnacles and other organisms (Stott 2003), for he never returned to the sea after his five-year voyage on the *Beagle*. We find that, for many ecologists, these transcendent moments when our personal affinities for our subject organisms and areas are fully expressed are the critical motivators that keep us going through the hours of tedious field and laboratory work.

Combining a broad use of our senses with a scientific ecological framework is a powerful and flexible application of observation-based ecology, allowing for a wide-ranging exploration of ecological systems, and also giving us the strength of robust ecological theory within which to fit our "new" observational knowledge. Observations, then, become part of a larger endeavor, where our personal experience can be translated into a deepened understanding of nature. This is not an easy process—it requires consistency and the ability to connect and communicate our findings, and to be able to test ecological questions scientifically (see Chapter 7). And given the challenges of environmental change and the opportunities for new observational tools now available, it increasingly requires us to move beyond our bodies and meld our natural observational senses with decidedly unnatural technological sensors, essentially creating a new symbiotic organism who we will meet in the next chapter—a modern technophilic naturalist-ecologist for the era of global change.

CHAPTER 4

# Using Technology to Expand Our Observational Senses

For all our unused sensory power discussed in the previous chapter, our senses are still decidedly limited, and this would restrict our ability to use an observational approach to ecology were it not for technology-facilitated expansions of our senses. This chapter is about how we can use technology to our advantage, how it can lead us astray, and how to harmonize the relationship between the biophilic observer of the natural world and the technophilic scientist who sees the world through technological sensors.

## Observing the Undiscovered and the Unexpected

Some of our sensory limitations reflect the fact that we humans lack certain observational equipment. This can hamper us at the very finest and the very largest scales of detailed observation. We can't send and receive signals in polarized light like cuttlefish (Brooks 2008), and we can't see the ozone layer thinning over the Antarctic. Sometimes our sensory equipment seems like the discount version of those that are so much more developed in other organisms. Like dogs, we smell with our nose, and the information is processed in a dense package of nerves known as the olfactory epithelium. But dogs have their olfactory nerves packed into an epithelium that is 17 times bigger than that in humans. It's no wonder smell plays a lesser and less conscious role in our observations, while it probably plays a dominant role in canine observations. We have evolved with a specific set of senses to survive in a specific environment, and those senses often fall short when we try to understand the complex world we live in.

But we don't have to settle for the sensory equipment we were born with or even the advanced sensory abilities we can develop through practice. Symbiotically combining observational natural history and technology—old and new—allows us to expand our observational toolbox. Indeed, our capacity to conceptualize, observe, and conduct science outside our normal sensory range has been well-established, as illustrated, for example, by the progress made in astronomy. But unlike today's astronomy, our ecological technologies do not have to be extremely "high tech." Sometimes just a slight technology-modified alteration of our sensory abilities is enough to spur startling new ecological discoveries. For example, an intensely curious evolutionary ecologist and avid diver, Nico Michiels from the University of Tübingen in Germany, made just such a discovery when he simply added a red filter to his dive mask. Instead of the expected complete blackout (because red light doesn't penetrate more than 15 meters or so in seawater), Michiels found an underwater world teeming with red signals—flashes of red eyes from fish and even whole fish bodies that appeared red. Theory alone would suggest that red would have little importance for fish at depth, but observations—followed by controlled experiments in Michiels's lab—are suggesting that fish use fluorescent red as a secret signaling device (Pain 2009).

In some cases, technologies we have had for centuries are becoming dramatically more powerful. For example, microscopy has advanced to three-dimensional imaging that can provide unprecedented visualization of the ecology of the immune system, revealing that immune cells and their targets interact and communicate in ways not unlike nerve cells (Davis 2006). High-speed video microscopy has uncovered the mysteries of how tiny copepods can ambush prey in a relatively viscous fluid environment (Kiorboe et al. 2009).

In other cases, the combination of new sensing technologies and the ability to step far back from our subject area by using airborne and spaceborne probes has provided us with a far more complete picture than our innate visual frequencies and field could provide. This technique is known as remote sensing and it plays into almost every field of ecology, from monitoring invasive species to determining nutrient flows to documenting responses of species to climate change. Moving away from the Earth's surface in order to observe has created a revolution in ecology and earth sciences. First, airborne equipment used mostly for military purposes gave us a much broader view of ecosystems. After World War I, geographers and ecologists started to use aerial photos to map plant communities and habitats and to understand patterns and processes from a scale that humans cannot grasp from the ground. This is reflected in Forman's pragmatic definition of landscape as "what one sees out the window of an airplane" (1995). Another leap forward was the popularization of satellite imagery, capturing reflections of Earth from the visual range to infrared and other specific wavelengths. Global ecological processes with important consequences for conservation have been quantified and understood by using satellite information. A major ecological and conservation story that began to unfold in the 1980s was the direct quantification, developed solely from satellite data, of deforestation rates in the Amazon and other tropical basins (Skole and Tucker 1993). More recently, such analyses have become predictive rather than just descriptive. Cassia Prates-Clark and her colleagues have used several types of remotely sensed data to more accurately predict the spatial occurrence of particularly vulnerable timber species across the Amazon Basin (Prates-Clark, Saatchi, and Agosti 2007). Now, radar technology in the form of LIDAR has opened new possibilities for revealing ecosystem structure at very fine scales from the air or space (Vierling et al. 2011; Vierling et al. 2008).

Climate change sciences have benefited immensely from remote sensing. Satellite information can help confirm data collected by ground-based weather stations, and it has a near continuous coverage across the Earth's whole surface. Understanding the spatial distribution of climate processes is a requisite for climate simulation, and without satellite data that is almost impossible. For example, Nemani and collaborators (2003) found that climate change has been changing cloud dynamics in the Amazon, which in turn may cause an increase in primary production in the area

(Arias et al. 2011). The sensor MODIS, which was launched aboard the satellites TERRA and AQUA in 1999 and 2002, respectively, allows us to take the pulse of the Earth without leaving our office. MODIS is now creating a massive amount of real-time data that can be applied to diverse ecological purposes, such as the detection of fires and changes in forest understory vegetation in areas affected by human activities (Morton et al. 2011).

Technology is also breaking down the dichotomy of humans as the observers and animals as the objects of study, as it allows us to experience the natural world through the movement and even sensory experiences of organisms in nature. Pop-off satellite tags, archival data recorders, and animal-borne cameras are effectively turning thousands of marine organisms into natural historians, providing unprecedented records of animal behavior and the marine environment (Moll et al. 2007; Block et al. 2005). In some cases, these animal-borne observations are wreaking havoc with our experimentally derived assumptions. Several marine mammals fitted with sensors, for example, dove far deeper and longer than any laboratory-derived physiological models suggested they could (Moll et al. 2007). In other cases, animal-borne observations are providing ways to answer difficult management questions that have gone unanswered for years due to lack of funds for necessary observations. For example, cameras fastened to seabirds, which have photographed both the feeding activities of the birds underwater and the human activity of boat traffic above water, were useful in demonstrating that fishing fleets and seabirds concentrated on different types of sardine patches when foraging, which has implications for how to protect and allocate scare sardine resources to both birds and people (Gremillet et al. 2010). And in other cases, animal observers are destroying the entire basis for our management structures. Barbara Block pioneered the use of satellite tags on marine predators, especially endangered bluefin tuna, which have been managed in the Atlantic as two separate stocks, divided by an imaginary line in the middle of the ocean. Block's tagged tuna have shown us definitively that they don't pay attention to human management boundaries at all, routinely mixing eastern and western Atlantic populations (Block et al. 2005; Block 2005).

Technologies can even reveal the subtle ways in which organisms perceive phenomena with their own senses. For example, ground squirrels use auditory calls to deter bird and mammal predators, but they use a different behavior—"tail flagging," in which they puff up and wave their tails to distract and confuse snakes, which do not hear. These responses to predation threat are easy enough for a human to observe, but it took infrared cameras to reveal that, while tail flagging, squirrels also heat their tails when signaling to pit vipers (which "see" in infrared), but not when signaling to gopher snakes, which lack heat-sensing ability (Rundus et al. 2007). Here, technology revealed an animal with defense mechanisms that are not just precisely tailored to their predators, but precisely tailored to their predators' observational senses.

## The Potential Pitfalls of Technology

Much of the technology we use in ecology has been developed for other purposes and then adapted to the needs of ecologists. Aerial photography, satellite imagery, and GPS all began as military tools. Gas-analysis instruments (e.g., LI-COR), first used in chemistry, are now used to measure photosynthesis. Auto-recording microsensors such as "iButtons" used by ecologists to measure temperatures experienced by organisms in the field (such as mussels on the rocky shore) were developed by the food industry to ensure that foods stayed at an optimal temperature as they traveled from warehouses to markets. And by the time a new technology is available to ecologists, it usually comes at a considerably reduced price. This is good news but also reflects that we as ecologists often have very little control over what new technology is being created; rather, we usually adapt our approaches to existing technologies.

Technology has continually helped ecologists see more and delve deeper into the complexity of ecological systems, but it is not without its problems. Technology can be costly, it has the propensity to fail, and it can lead us into a world full of meaningless data. Some problems, especially the human bias to "see what we want to see," have always plagued the interface of observational and technology. Using the earliest microscopes, biologists of the day were convinced they could perceive small homunculi—miniature humans—on the heads of sperm. Others are wholly novel problems of our era.

We see four major classes of problems that need to be considered when incorporating technological observations into ecological science. These are problems specifically related to the use of technology, although they mirror some of the larger methodological and philosophical issues that we discuss in Part III of this book. First, there are issues concerning the unequal availability of technology. Second, there are legitimate concerns about the long-term accessibility of observational data collected with new technologies and recorded digitally on ever-changing storage media. Third, there is a fundamental tendency to draw spurious conclusions from data seen through a technological filter. Fourth, there is the simple but profound problem of technologies putting up another barrier, another screen, between ourselves and nature. Here we present examples of these four kinds of problems along with suggestions of how we might avoid them as we move forward, inevitably, into an ecology that is both carbon- and silicon-based.

Our first concern is that technology is not readily available to all ecologists in every part of the world, especially to those in developing countries. Even satellite information is not collected or made easily available in some areas of the world. Lack of funding for extensive ground-data collection can be somewhat ameliorated through technology that allows more data collection with limited resources (e.g., remote sensing, rapid field-assessment technology such as video recording or photography). But even then, collected and analyzed data may remain inaccessible. In particular, a lot of the data that could be used in meta-analyses, which are increasingly important for comparing ecological dynamics across large scales (See Chapter 6), are locked up in published papers in very expensive scientific journals. While there is an increasing movement toward "open access" journals, far too many articles—even those resulting from publicly funded research—are inaccessible to far too many ecologists. As a result of these external and internal forces, developing countries need to rely on data donations and may lack the technical capacities for data analysis.

Fortunately, as scientists and politicians realize that ecological problems have no borders, more and more data is becoming available at no cost for the world's scientific community. The use of Internet technologies and networks of scientists (see Chapter 6) that can facilitate international collaboration can also help alleviate some of the accessibility problems. As cell phones become ubiquitous even in remote areas of developing nations, new freeware applications such as CyberTracker (cybertracker.org) and Ushahidi (ushahidi.com) are creating huge communities of citizen-scientists compiling and sharing social and ecological data. Still, much more needs to be done in order to make adequate use of this information for scientific and applied purposes in developing countries.

Second, although it hasn't been formally studied, the increasing rate of change in technology and the rapid obsolescence of equipment and software, as well as a lack of adequate training and personnel, can complicate the use of technology in ecology. A clear example is the application of the spatial techniques available for ecological analyses, which are rarely used by ecologists who have no specialization in this method, limiting its use to a small proportion of researchers. Ecologist Thorsten Weigand has tried to set an example of how to move beyond this problem by developing an open-access spatial-analysis software called Programita, which he updates and tests with each successive iteration of his spatial ecology course.

Kristin Wisneski, a University of Arizona graduate student who has been developing and testing smartphone applications to help get young students interested in field-based ecology and natural history, has found that keeping programs up to date and helping teachers learn the technology require a tremendous investment of time. Educators in the field experience their own set of challenges, too. While mobile and location technologies (such as GPS) facilitate the capture, storage, and sharing of observational data, they pose great challenges in formal and informal learning situations. Social media, music and video smartphone applications, and full-time access to an Internet browser bring distractions into the classroom. Some teachers, though, have already begun exploring those very distractions as means of motivating and inspiring students by, for example, helping them develop youth-driven projects that integrate informal learning and the use of technology in order to create linkages between the classroom and the natural world.

An emerging worry about the ongoing availability of high technology is the impermanence of computer-stored data. Certain satellite data are already becoming impossible for scientists to access because of the inability of government agencies in charge of data to maintain databases stored on outdated digital platforms (Loarie, Joppa, and Pimm 2007). More frightening are estimates of the rapid degradation and loss of electronic files. Damage to storage media and the files stored on them, as well as software incompatibility and differences in how "metadata" (data about the data, like the dates on which data were saved or modified) are transferred between electronic media, all result in potential losses, even of relatively recent data. For example, Brad Reagan (2009) notes that an electronic updating of William the Conqueror's "Domesday Book" (which documented daily life in Britain in the eleventh century), compiled by the BBC in the mid-1980s, is now less accessible, due to file incompatibility, than the original Domesday Book written on parchment in 1086! These kind of incompatibilities put an extra premium on getting the metadata well documented and backed up. As we will discuss in Chapter 6, well-coordinated scientific networks offer opportunities to address some of these issues by developing mutually reinforcing incentives to harmonize data collection, storage, and dissemination across the globe.

Third, the larger problem with technology is that even the best technological sensors can lead to completely spurious conclusions about the underlying ecological phenomena. This is not a problem of inadequate technological progress. Although remote-sensing data are in some cases still too coarse-grained to fully document some key ecological phenomena (Herrick and Sarukhan 2007), history strongly suggests that technology will improve to fill these gaps and provide unexpected new tools. Indeed, all technological approaches, whether they use molecular genetics or remotely gathered data, must be backed by solid natural history to avoid misinterpretation. Technology gives us one very narrow perspective about a particular ecological phenomenon, and we should be particularly careful about overgeneralizing results, mindful that our conclusions may change with the temporal and spatial scale of observation, the study area, and the specific phenomena we are observing.

Several years ago, alarmed by the loss of coastal mangrove forests due to tourism development, Exequiel Ezcurra and colleagues calculated the potential dollar value of lost mangrove forests in Mexico due to the loss of fisheries, nurseries, and habitat for other key elements in coastal food webs (Aburto-Oropeza et al. 2008). In response, some researchers, using satellite-data analysis showing that the extent of mangrove forests in Mexico had actually increased, argued that the economic loss was overestimated. Relying solely on the infrared satellite images of the coast, it was indeed clear that the extent of mangrove forests had increased. But when Exequiel and colleagues arrived on the ground to look at the actual mangroves that had supposedly increased, they found a much more complex picture in which climate warming, sea-level rise, and habitat destruction were all at work. What happened was that higher sea levels—likely caused by climate warming—as well as flooding during El Niño events and tropical storms, led to greater inundation of the mudflats behind coastal mangrove forests (Lopez-Medellin et al. 2011). This meant that mangrove propagules, which grow best in a wet environment, could occasionally take root back behind a mangrove forest during the temporary flooding. The problem is that although these pioneers added to the overall extent of mangrove forest seen in satellite data, they provided little service to the ecosystem as habitats for organisms or fish nursery habitat because they lived on mudflats that were rarely underwater. Similar controversies have been sparked with regard to forest recovery and degradation in other regions such as the Pacific Northwest of the United States, where secondary-growth trees are now old enough to be classified as forests from remote observations, leading to the conclusion that the total forest cover has increased, while the ecologically richer primary forest cover has actually decreased (DellaSala 2011). Cases like this, which are sharp warnings about the perils of becoming too dependent on technology, are also pointed reminders not to become too narrow in our approaches. They remind us that there is value to becoming an ecologist who intimately understands natural history and natural relationships, no matter what methodological approach—technological, experimental, or theoretical—is ultimately used.

This brings us to the fourth, and less quantifiable (but not trivial) concern that too much technology, even if operated in service to nature, puts us at too great a remove from nature itself. Nature is not just the source of the "raw data" to an ecologist, but a source for inspiration, creative thinking, and regeneration, all of which are as necessary as sampling and statistical skills for being a productive ecologist. It's easy for us to lament the decline in "forest hours" among children and their exponential rise in "screen time," but adults can suffer the same maladies by spending too much time in front of our own screens. We may be tempted to use every new technological gadget just because it is available, but we need to give adequate thought to the specific questions this new technology will help us answer. Field observations and a deep understanding of our study systems are crucial before we become obsessed with new technology. Like many symbiotic relationships that become permanent, the relationship between natural history and our technology must balance benefits and costs, and this balancing act may fundamentally change the distinguishing characteristics of an ecologist. When combined with the kind of deep, natural history–based understanding that comes through careful observation, these technologies can, ideally, give rise to a new kind of ecologist, whom physiological ecologist Carlos Martinez del Rio of the University of Wyoming calls an "ecological cyborg." In his words,

> Contemporary natural history is for cyborgs: creatures simultaneously human and machine. The distinction between a naturalist cyborg and just a cyborg depends on the traits that have been traditionally associated with natural history, and which include finely honed ethical and esthetic instincts, biophilia, and the observational powers and intuition that result from long hours in the field. (Martinez del Rio 2009)

Although Martinez deftly unites several important aspects of modern ecology here, our only reservation is that, for most people, the word "cyborg" invokes something more machine than human, while we view ecology as an inherently people-oriented science. The relationship between human systems and ecological systems were prominent areas of study for ecologists of the early twentieth century, operating between two devastating world wars. Curiously, though, much of the postwar twentieth-century ecology ignored humans and human impacts, instead conducting work in pristine laboratories or in scientific nature reserves where the normal activities of humans are assumed to be nonexistent. Human behavior, human decision-making, and human psychology are rarely used as variables in ecological studies even though these are the most significant variables in most ecological systems. But humans, beyond just the human ecologists conducting studies, are beginning to play a central role in ecology once again, both because of their undeniable impact on almost all ecological systems and because of the contributions that all kinds of people—not just scientists—are making to ecological science. In the next chapter, we discuss these contributions, how they are expanding what we can study in ecology, and how they are forcing us to ask new questions about the interface of science and society.

CHAPTER 5

# Local, Traditional, and Accidental Ecological Observers and Observations

One of the most notable features of an observation-driven approach to ecology is that *data can come from anywhere*. There are virtually no limits on the types of observations that might become part of a scientific study of changing ecological systems. Old photographs, a naturalist's field notebook, seafood-restaurant menus from a bygone era, long-forgotten scientific papers, a gambling contest, feathers of a bird preserved in a museum, stories passed down from generation to generation, and even a centuries-old pack-rat midden preserved by generations of pack-rat urine have all been used recently in ecological studies. This openness is both a benefit—it creates limitless opportunity for ecological studies and also invites all sorts of people to become part of a new ecological understanding, regardless of their scientific training, means, or geographic location—and also a curse—how do we sift through it all to find out what is useful, and once we find what we are looking for, how much can we trust all these uncontrolled observers?

With this open view of ecological data, the high-tech wizardry we gushed over in the last chapter is put into proper perspective as just one means of achieving a larger ecological understanding. Some of the best observations of nature come from people who have little or no technology at their disposal. This chapter is about humans who have observed the environment closely for long periods of time and passed these observations down through generations, and about what they can contribute to scientific ecology.

### Ecological Knowledge from Local and Traditional Observers

Humans have developed a number of ecological observing systems that rely on both their innate senses (Chapter 3) and culture, rather than technology, to transmit and improve the accuracy and utility of their findings. Moments before the devastating 2011 tsunami in Japan, for example, fishermen who were out to sea and felt the trembling of an earthquake remembered their grandfathers' observations that "tsunami do not rise in deep water," and quickly stopped fishing and moved further offshore, letting the tsunami wave gently pass under them (Shimbun 2011).

Although it may have been first developed as a survival mechanism, culturally transmitted ecological knowledge is not limited to ecological concerns of immediate relevance to our survival. Much of it arose as part of the intergenerational preservation of cultural identities expressed in artwork and mythology built around natural organisms or natural phenomena (Dayton and Sala 2001). Other data has long been amassed for the sake of maintaining both the seasonal and long-term ability to harvest plants and animals (Fleischner 2005). These types of data and practices have come to be known as Local Ecological Knowledge (LEK) and Traditional Ecological Knowledge (TEK), the latter of which has been defined by the Ecological Society of America as "adaptive ecological knowledge developed through an intimate reciprocal relationship between a group of people and a particular place over time" (www.esa.org/tek). LEK differs somewhat from TEK in that it does not necessarily require a tradition of knowledge handed down through generations (Gilchrist, Mallory, and Merkel 2005). A first-generation fisherman, for example, may acquire excellent local knowledge of the local ecology through the course of her life's work.

Often these nature observations have been made as a routine part of daily life. Traditional and local knowledge holders have spent abundant time in direct connection with nature—as ranchers and farmers, herbalists and artisans, fishermen and foresters. Through this connection they are able to observe far more and with far greater context than a scientist might in a limited field season.

In fact, the beginnings of science were marked by the systematization of traditional observations. As natural sciences evolved, there developed a division based on the authority of observers (qualified vs. amateur) that ultimately devalued TEK and LEK. Nonetheless, scientists have long been aware of the value of traditional knowledge in gaining ecological understanding.

Consider again Samuel Lockwood's "Something About Crabs" paper from *The American Naturalist* in which he noted, "We knew some years ago an old crabber, wholly illiterate, but whose intelligence was above average. . . . Often when supplying the family with fish, has he been closely questioned by us about the crabs. . . . " (Lockwood 1869) There is a clear note of condescension discernable in Lockwood's account and this reflects the treatment of traditional and local ecological knowledge holders and their data for much of the history of ecological science. Nonscientists were commoners whose charming anecdotes might add some color to a scientific exploration but could never achieve the precision or depth of understanding attainable by a man of letters.

But science is increasingly recognizing the value of data gathered by nonscientists, a trend that Gary Nabhan has tracked and discusses in Box 5.1. This new wider acceptance is due in part to a new-found respect for nonacademic knowledge and non-Western lifestyles, but it is also purely pragmatic—the halls of science have failed over the last century to produce much of the data we would like to have if we are to understand our changing planet (Dayton 2003).

## Valuing the Role of Different Observers in Ecology

The large gaps in ecological information have opened the door for the acceptance of data that otherwise would have been dismissed. Informal and traditional cultural sources of data are filling these gaps and being used in ecological science and conservation management. In his paper "The Case for Data-less Marine Resource Management: Examples from Tropical Nearshore Finfisheries"—a title likely to have rankled

BOX 5.1

**Traditional Ecological Knowledge and Observation-Based Ecology**

GARY NABHAN

The traditional ecological knowledge of place-based indigenous and peasant cultures is perhaps the oldest fund of observational ecology or natural history data extant on this planet. However, because most of it has been orally transmitted over the decades and centuries, most Western scientists have had neither access to it nor much respect for its value. That is ironic, for the seeds of modern ethnobiology were planted at about the same time as those of modern ecology, between the 1860s and 1890s; since that time, an ample and often insightful literature of traditional ecological literature has emerged that should rightly be studied and celebrated by all observational ecologists, regardless of their cultural origins.

Contrary to popular misconception, ethnobiology demonstrates that traditional knowledge of plants and animals extends far beyond their names in indigenous languages and their uses by subsistence cultures. In fact, it can be argued that indigenous ecological knowledge includes true intellectual inquiry into the relationships among plants, animals, cultures, and their habitats rather than merely being a utilitarian resource-management practice. Traditional ecological knowledge embraces the domains of biosystematics, anatomy, physiology, phenology, chemical ecology, community ecology, and even agro-ecology. It, like much of observational ecology and natural history, analyzes "found experiments" such as lands differentially affected by wildfires, floods, weather shifts, or grazing to infer ecological effects or certain "treatments." In its interpretations, traditional ecological knowledge is certainly imbued with orally transmitted knowledge from other generations, so it often discerns cause-and-effect relationships that do not necessarily reflect from observations made exclusively during just one lifetime. It typically uses no statistical methods to discern longitudinal patterns, but does intuitively integrate many observations taken over long periods of time.

professional fisheries managers who rely on reams of data and complex mathematical models to determine "optimal yield" targets for fishing efforts—R. E. Johannes outlined cases from South Pacific cultures where traditional knowledge regarding fishing practices was at least as effective as quantitative "scientific" management (Johannes 1998).

As such, it harbors within its fund of knowledge observations that could have not been made by Western-trained scientists who have more recently arrived in the same habitats.

For instance, Seri Indian elders recall observations of California condors on Tiburon Island in the midriff of the Gulf of California from the 1920s and early 1930s. Hopi elders also orally transmit similar observations of condors made by their ancestors, observations that predate condor extirpation in the Grand Canyon around the 1890s. Such orally transmitted data points derived from traditional ecological knowledge may be used to broaden or refine options for the recovery of endangered species.

Similarly, the Seri have already played a role in the recovery of endangered sea turtles through providing conservation biologists with an extraordinary set of data on leatherback turtle nesting-beach location and feeding-ground location, as well as their diet and behavior. The probability of conservation biologists independently coming upon such a wealth of observations regarding this rare marine reptile—given the species' low population density and level of endangerment—is exceedingly low.

And yet, the complexities of how the Seri use their native language and Spanish to communicate such knowledge remains a barrier to fully integrating such knowledge into species-recovery plans. It takes someone conversant in their language and familiar with sea turtle biology to fathom the depth and utility of the Seri's understanding of leatherbacks. Such depth cannot be "extracted" or "downloaded" in a single interview or even a single season.

If observational ecologist or natural historian is the world's oldest profession, it is also among the most endangered of professions due to language loss, oppression, and economic domination of indigenous peoples. Nevertheless, the indigenous youth of today are surely among the rightful heirs to the legacy of natural history as a cultural practice.

Hundreds of scientific papers in the last decade discuss TEK and LEK, and although scientific ecology was slow to admit them into actual research projects (Gilchrist and Mallory 2007), papers are increasingly using these forms of knowledge to address subjects as varied as climate change's effects on phenology (the timing of natural events like flowering or hibernation), harvesting effects on natural populations, food-web structure, and migration patterns (Salomon, Nick M. Tanape, and Huntington 2007). For example, a March 2011 search on the Web of Science revealed 275 papers that cited the seminal paper "Rediscovery of traditional ecological knowledge as adaptive management" (2000), which introduced much of the ecological science community to the value of TEK and the prospects for incorporating it into their research. Most of these papers present examples in which TEK or LEK was used to gain ecological insight or was applied to management issues. Jeffrey Herrick and colleagues have shown that national ecosystem assessments, with outcomes suitable for affecting management and policy decisions, can be developed by combining ground observations, remote sensing, and cellphone data recording with LEK and more quantitative scientific observations (Herrick et al. 2010). The Ecological Society of America now has an active section on Traditional Ecological Knowledge that aims to foster the respectful use of TEK as well as to encourage more active participation of indigenous people in ecological science. In other words, noninstitutional forms of knowledge are becoming institutionalized.

At the same time, in a world increasingly populated by observant humans, some important data for understanding ecological dynamics are collected with no thought beyond their immediate use. These might be past scientific studies aimed at particular narrow questions, or what we might call "AEK" (Accidental Ecological Knowledge)—information that only later was discovered to have ecological significance. For instance, a photo of the Swiss Alps may have been created for a tourist postcard but now is an important data point in a worldwide study of retreating glaciers (Webb, Boyer, and Turner 2010). Rafe analyzed data from each year of an 87-year-long series on the exact time in spring when the ice in the Tanana River in Alaska melted and found that trends in the timing of spring melt coincided closely with both long-term climate warming and multi-decadal variations in warm and cool periods throughout the twentieth century (Sagarin and Micheli 2001). The source of these data wasn't a long-term scientific study but an ongoing gambling contest in which participants had to guess the exact minute in which a wooden tripod put out on the ice in winter would fall through the thawing spring ice. With $300,000 on the line and hundreds of people camped out on the riverbanks in anticipation of the moment when the tripod falls, the overall record is both the most accurate record of spring ice melt (many observers are watching and ensuring that the time is correctly documented) and the most precise record (it is recorded down to the minute, rather than day, or week, of melt).

In this case, the value of the data doesn't come from any particular local wisdom (the vast majority of participants in the ice melt contest guess the wrong time), but from simply being the only reliable data source available. While there are some long-term scientific temperature records from interior Alaska, they suffer from large temporal gaps in recordings, known inaccuracies, and frequent movement of weather stations (which exposed recording equipment to very different microclimates at different points in the record). For these reasons, it has been noted, ironically, that ice-melt records may be "more accurate long-term indices of air temperature than air-temperature records themselves" (Assel and Robertson 1995).

## The Limits of LEK, TEK, and AEK

Local knowledge can at times be too narrowly focused. Local resource users may have unparalleled knowledge of the species they harvest and some of their immediate ecological relationships, but if they group all other species into a less important category, then the quality and quantity of available knowledge may not match what is needed for an ecological study. Indeed, the narrowing of traditional ecological knowledge likely began, long before recorded history, when we first became agriculturalists and needed to cultivate more-specialized knowledge about particular species and ecological phenomena (Fleischner 2005).

Data sources in historical ecological studies also inevitably suffer from being too narrow in scope. Whether taken accidentally or deliberately as part of a scientific investigation, historical data sources almost never have as much spatial or temporal resolution, taxonomic diversity, or detailed description of the "metadata" (the information about how and why the data were taken) as we would like (Sagarin 2001).

The stability and utility of informal ecological knowledge rests, sometimes precariously, with the individual knowledge holder and with the cultural context in which the knowledge is generated and held. For example, although informal ecological knowledge has been important in evaluating the effects of climate change, some of these forms of knowledge may be unable to keep up with rapid changes to social and ecological systems. Johannes, for example, showed that generations of South Pacific fishing communities thrived with "data-less" management, but he acknowledged that more quantitative data and scientific methods will be necessary to continue sustainable management in the current era of rapid change (Johannes 1998).

As with all forms of data, the accuracy of informal knowledge can degrade over time. Two groups of researchers working in different parts of the Gulf of California found that younger generations of fishermen considered the maximum size of any particular fish species to be significantly smaller than fishermen from older generations (who had personally seen much larger fish), and younger fishers were also much more likely than older fishers to think that little change had occurred in their fishery (Lozano-Montes, Pitcher, and Haggan 2008; Saenz-Arroyo et al. 2005). But these studies also showed the value of learning from the knowledge of people with a long history of interaction with nature, before their knowledge is gone forever. Both the passing of individual observers and the more troubling passing of entire cultures and language groups have long caused concern among anthropologists, but this should be a concern to ecologists as well (Davis 2010).

How do traditional sources of knowledge stack up against more mainstream data sources? Local and traditional methods of ecological observation can be compared directly to more mainstream methodologies when they are both used in the same study. Gilcrest and colleagues evaluated LEK of Inuit people against scientific studies related to populations and distributions of four marine bird species and found a full range of accuracy, from low to very high (Gilchrist, Mallory, and Merkel 2005). They conclude that LEK can be an essential complement to more mainstream means of achieving ecological understanding, but that it can rarely stand on its own as a basis for guiding conservation management in an era of rapid and widespread ecological change. At the same time, Brook and McLachlan caution that when we compare mainstream "scientific" data and informal data sources we should avoid the preconception that data collected in a typical scientific design create the "correct" baseline against which to test the "suspect" traditional or local knowledge—indeed, several cases have been found in which LEK provided more-accurate pictures of ecological cases than did scientific studies (Brook and McLachlan 2005).

Indeed, the benefits of using traditional knowledge may be overlooked if we only consider what gets published in the "Results" section of a typical scientific paper. Attum and colleagues, for example, tested the efficacy of experienced human observers against radio telemetry for tracking endangered tortoises in Egypt. They found that both methods yielded similar results in terms of efficiency and accuracy, but employing human observers provided the added benefits of creating incentives for conservation and greater interest in the long-term research goals than if the scientists had simply used remote sensing with radio tracking (Attum et al. 2008).

The incongruence in conclusions based on different sources of data highlights a key point about the fallibility of any human observers. Even the most admired observers of nature have sometimes drawn incomplete or wholly inaccurate conclusions about the natural world, hobbled by their inability to see across large spans of time or space. John Muir, a keen observer of the world who could convey the grand scope of California's Sierra mountains and valleys to generations of readers, and who even witnessed the destruction by dam of his beloved Hetch Hetchy Valley, nonetheless grossly underestimated the ability of humans to alter natural systems, writing, "Fortunately, Nature has a few big places beyond man's power to spoil—the ocean, the two icy ends of the globe, and the Grand Canyon" (Muir 1918). In struggling through the figures for their next edition of the Pacific Coast field guide *Between Pacific Tides*, Ed Ricketts confided his confusion to his co-author Joel Hedgpeth about the average ocean temperature data (isotherms) that he was receiving from the Scripps oceanographer Harald Sverdrup, which didn't coincide with previous records. "The only other explanation is that the isotherms have changed," he wrote. "This is known to have happened in the past, but I always thought of that in terms of the geological past. If not even the mean isotherms are going to stand stationary long enough for them to be standardized, where are we?"* And although Ricketts consciously sought to achieve a much larger view of ecological understanding—what he called the "toto picture"—by his own admission he often fell short of that goal. Along with his friend John Steinbeck, he wrote of their disappointment, after an eight-week journey through the Gulf of California, that they "could not yet relate the microcosm of the Gulf to the macrocosm of the sea" (Steinbeck and Ricketts 1941), meaning that despite all their intensive observation, the connection between their relatively small body of water and the larger oceanic ecosystems was still not clear.

The passage of time, the compounded impacts of human activities, and new technologies developed by humans have both revealed the limits of these earlier observations and their importance as baselines for marking how much has changed. Muir would be horrified to learn—although it is common knowledge to almost all ecology students today—that the Grand Canyon, the oceans, and the poles have all been radically and irrevocably altered by humans. By contrast, we are equally struck to know that less than 100 years ago such impacts could not even be predicted by the most ardent environmentalist. It would only be a few decades after Ricketts's death in 1948, through the use of observational data taken as part of a long-term oceanographic monitoring program developed at Scripps, that we would learn that the mean isotherms of ocean temperatures were, in fact, changing rapidly, and that the most likely agent of change was human-caused climate warming (McGowan, Cayan, and Dorman 1998; McGowan 1990). Yet these changes could never have been discovered without the prescient vision of Ricketts, who publicly called for a long-term oceanographic monitoring program (Ricketts 1945–47), and the Scripps scientists who made such a vision a reality. And when Rafe

*Ed Ricketts, letter to Joel Hedgpeth, 9 December 1945. Edward Flanders Ricketts Papers, 1936–1979. Special Collections M0291, Stanford University Libraries, Department of Special Collections and University Archives.

and several other scientists returned in 2004 to the same locations studied by Ricketts and Steinbeck in 1940 to document changes to the Gulf, the microcosm and the macrocosm that Ricketts failed to connect could easily be reconciled in observations of coastal development, emerging zoonotic diseases, overfishing, loss of top predators, and climate warming, all of which have altered the Gulf and many parts of the world's seas in parallel (Sagarin et al. 2008). Yet this connection could not have been made without the detailed and insightful literary record laid down by Ricketts and Steinbeck.

The promise and pitfalls of informal ecological observations are an intensified reflection of those facing all scientific data. We believe that the benefits of a much more inclusive ecological science far outweigh the costs, both in terms of their present value and in terms of the positive feedback cycle that is generated by getting more people and perspectives involved in observing nature and its changes, developing or validating their own sense of biophilia, and this in turn stimulating a desire to protect and restore natural systems. Moreover, both mainstream scientific investigations and local ecological knowledge can be improved with reference to one another, and likewise both scientists and local communities can benefit from mutual sharing of ecological observations. Opening up ecological science so widely is already creating a rush of new data. Nonetheless, these data aren't evenly available everywhere and to everyone, and their quality varies widely. In the next chapter, we show how the continued success of observational approaches depends critically on how we create, analyze, share, and care for observational data.

PART III

# THE CHALLENGES POSED BY AN OBSERVATIONAL APPROACH

In this part of the book we step back from our unbridled enthusiasm for observational approaches to look more soberly at all the difficulty they cause. There are a number of practical difficulties of dealing with sources of data that can be woefully sparse at the times and in the places we'd most like to have them, and yet in other times and places they can be so frighteningly abundant that not even our biggest hard drives can handle them. In Chapter 6 we illustrate some solutions to data problems that involve both a renewed valuation of old data-storage methods, such as natural history museums, and heightened attention to new means of analyzing and sharing large observational data sets.

But even given the challenges of finding, storing and analyzing ecological data, there are still fundamental questions that arise concerning observational approaches to ecology. It's not the idea of observing *per se* that it is in question, but how observations are to be used, that has caused longstanding scientific and philosophical debates. In Chapter 7 we get right to the heart of these debates with a fundamental question: "Are observational approaches to ecology scientific?" In exploring this question, we raise the most common critiques of observational approaches and show why these critiques are far less devastating to observation-based arguments (and sometimes wholly irrelevant) in an era of massive observational data pools that can be examined and cross-examined using the type of analyses we discuss in the following two chapters. Ultimately, our answer to the fundamental question is "yes, observational approaches to ecology *are* scientific"—but this is tempered by the knowledge that there is a high bar to making observations that are both scientific and useful to science.

CHAPTER 6

# Dealing with Too Many Observations, and Too Few

In the last chapter we showed that a greater openness to the observations of nonscientists is unveiling valuable new data sources and even accidental ecological knowledge. In this chapter we focus on the more formalized types of observational data that ecologists have been taking for well over a century in the form of museum collections, historical data, long-term monitoring schemes, and more recently, networks of ecological observers. How ecologists plan to collect these data, how the collections or observations are maintained and stored over long time periods, and how they are analyzed all ultimately affect the strength of the conclusions we can draw from them.

Ecologists have always used unmanipulated observations, and some classic ecological texts from decades ago, such as Jared Diamond and Ted Case's *Community Ecology* (1986) and Jim Brown's *Macroecology* (1995), highlight the value of large observational data sets. What is different now is that the sheer volume of observational data, the diversity of its sources, and its variability in availability and quality is unprecedented. Sometimes, as is the case within genetics and genomics, we are literally creating more data than can be stored, even on electronic media (Pollack 2011). At other times this situation leads, frustratingly, to a lot of simply useless data just lying around or accumulating on hard drives. And sometimes we uncover good sources of data, but there just aren't enough of them. We're facing a modern paradox, which is that even as we are flooded by some data, there are still many regions of the world and many ecological questions that are debilitatingly data-poor. Solving this paradox will mean addressing four issues: (1) identifying useful data that can be efficiently extracted from a much larger matrix of ecological observations; (2) finding ways to fill in gaps in historical, modern, and future data-collection efforts; (3) connecting observers and observations in order to achieve a global understanding of ecological phenomena; and (4) capitalizing on the myriad ways—old and new—of analyzing these data in order to put the data flood to beneficial use. The payoff from filtering through these data and cleverly layering multiple methods of data analysis is a remarkable array of studies that stretch the scope of ecology, giving us deeper insight into long-standing questions and new views of the changing world.

## Identifying Useful Data within a Flood of Data

Data can be collected in a multitude of ways for every specific question in ecology. Put a bunch of ecologists in one room with a question and ask them to come up with a list of data types that would be relevant to answering the question, and in a few hours you will have more ideas about what data is important to collect that you can afford to fund with the entire budget for ecological research. This is not trivial in an era where funding is limited and the unanswered questions are unlimited. Prioritization of which observations are most relevant is a difficult but essential task, not to be undertaken lightly.

We might think we'd like to have an infinite set of variables recorded over long periods of times to explore all potential relationships in ecological problems. Setting aside for now the enormous problem of how we might deal with all that data, the reality is that the data we actually can have depends on a combination on very uncontrolled factors (e.g., a volcanic eruption sterilizes all life on a mountainside, providing an ideal natural laboratory to study succession) and also the changing interests of the research community (e.g., research trends, funding priorities). Moreover, especially for long-term data sets, what data to collect and how to collect them are often decisions made several decades ago, long before the actual analysis of the data. By the time we want to test a particular question with the data, we may discover that they weren't aimed at addressing questions that are currently of interest, or they may have been collected in outmoded ways that leave large gaps in our understanding.

For example, data on fecal indicator bacteria (FIB) in coastal water bodies, which are relevant to many ecological questions (especially as we try to protect or restore water quality and habitat in coastal marshes), have been collected under various legally mandated frameworks in the United States. The data are collected at certain frequencies (e.g., daily or weekly) using simple bacterial-culture protocols that have been largely unchanged for decades. For an ecologist looking to study the effects of, say, a wetlands-restoration project on water pollutants, these data may be useful, but they aren't likely to cover the rapid daily shifts we now know occur in FIB concentrations (Dorsey et al. 2010), nor will they give us genetic information (which is now available through new molecular methods (Sagarin et al. 2009) on the likely source of those FIBs. (Did they come from the birds enjoying the restored wetlands, dogs being taken for a walk on the wetland nature trails, or from human sewage?) With political pressure to get the wetlands restoration going as soon as possible, but with the inertia of a deeply entrenched bureaucracy slowing any proposed changes to sampling protocols, there likely won't be time to implement the latest FIB assessment techniques before the restoration efforts in an ideal "Before-After Control-Impact" (BACI) study (Schmitt and Osenberg 1996). There will always be such mismatches in what we have and what we would like to have, in part because the time frames for technology advancement, political change, and management change are so different from one another.

Having a sense from the outset of what makes a good historical data set to use as a baseline can be helpful in both screening potential data sets for their usefulness and in planning effective new data-collection efforts, as Julie Lockwood shows by using historical bird atlases in Box 6.1. In the mid-1990s, as a few dozen papers appeared that linked climate change to changes in species distributions and populations, Rafe developed a checklist of data attributes that would be most useful for these kinds of

**BOX 6.1**

## Historical Records Shed Light on Biological Invasions

JULIE LOCKWOOD

*History teaches everything, including the future.*
—Alphonse de Lamartine

When students enter graduate school they often assume that the data they use in their dissertations must be hand-collected by themselves alone. Sometimes the better route is for each student to stitch together various data sources into one coherent long-term view of a research topic. A critical resource for such a student is historical records.

Historical records can include anything previously published in books, journals, or reports. The data gathered from such sources range from purely qualitative descriptions to tomes of quantitative data organized in tables and graphs. Although there are limitations to what can be inferred based on historical data, such information is amenable for use by the modern biologist and can be treated in much the same statistical way as one would treat self-collected data.

Along with my colleagues Tim Blackburn and Phillip Cassey, I have made much use of a particularly comprehensive and compelling historical record on birds introduced as non-natives to locations around the world. The data primarily come from two published sources. The first is a classic book by George M. Thomson that details the history of plant and animal introductions into New Zealand (Thomson 1922). Thomson collated the myriad sources of data on birds introduced to New Zealand via the activities of acclimatization societies, which were groups of citizens that devoted their time and money to importing and releasing

studies (Sagarin 2001). The strongest studies had some combination of good spatial resolution (a large number of field sites spread across a large area), good temporal resolution (studies spanning a large range of time, especially when data are taken frequently across that span) and good taxonomic resolution (sampling of a lot of different types of species so that one can see if there are differences in the effects on species with differing physiological needs, such as reptiles and mammals). Not all studies need birds (and other species) in order to satisfy aesthetic and practical goals. He created a rich source of information on the role of numbers of individuals released (propagule pressure) on non-native species' establishment success, among other currently relevant topics. Through a series of publications, my colleagues and I have shown the power of propagule pressure to explain variation in invasion success, across species and locations (Cassey et al. 2004).

The second is perhaps a more impressive collection of detailed records on bird introductions by John Long (Long 1981). Long completed a massive review of available literature to detail which birds have been introduced outside of their non-native ranges, which of those were successful in establishing self-sustaining populations (as of 1981), and the circumstances surrounding those introductions. Long's book has become a staple for a variety of investigations into the causes and consequences of non-native species establishment, post-invasion evolution, the extent to which interspecific competition structures community membership, and the role of invasive species in causing the extinction of native species. For example, the data in Long provided a critical resource for our comprehensive review of how avian invasions can inform basic theories in ecology and evolution (Blackburn, Lockwood, and Cassey 2009).

Historical documents contain a wealth of information relevant to invasion biology and other fields, but this source of data is vastly underutilized. In terms of our understanding of non-native birds, the historical details surrounding each species' introduction into a novel area are pertinent for understanding their subsequent evolutionary trajectory, interactions with co-occurring species, and population dynamics. Combining such historical information with the tools and techniques of modern biology can thus provide unprecedented insights into ecology and evolution.

to cover all of these areas (it would be unrealistic to think any individual study could) but data sets lacking in all of these areas may not be worth the effort and expense to analyze or maintain. In choosing which data sets to seek out, or to create, it helps to have a larger context in which to understand where a particular study would fit within the larger question we are trying to address. In the example here, the larger context is how living things are responding to present-day climate change—a context within which many different types of studies could be fit, each with its own strengths and weaknesses. Thus, even though each study in the early years of uncovering species responses to climate change may have only been able to confidently illustrate a small fragment of the story, when taken together the studies became a mosaic that clearly reveals a picture of whole communities being altered by climate change.

In general, to understand an ecological system, which operates by its nature at multiple scales, information should be collected at multiple scales to see whether patterns observed or experimentally determined are replicated at larger spatial or temporal scales, and to see whether certain patterns emerge only when viewed at large-enough scales. For example, to study biological invasions, observational data that come from multiple sources, from county weed records to national plant inventories, can be particularly useful for detecting patterns across multiple scales. Tom Stohlgren and his colleagues have been studying plant invasions in North America and around the world, and have found striking patterns that have challenged the results from small-scale experiments, as he describes in Box 6.2. More recently, using global herbarium data from 13 regions worldwide, Stohlgren et al. (2011) detected that a large proportion of the most widely distributed plants (those with a larger distributional range) are non-native, which highlights the fact that invasions are contributing to biotic homogenization at regional and global scales.

## Long-Term Data Storage to Prevent Data Droughts

Sometimes you don't get the data you want because it just isn't there. There are a lot of ecological data that haven't been taken yet, both in a trivial sense (because the total volume of ecological information to be probed is essentially infinite), and also a more pragmatic sense (because going out and recording natural observations has never been a top priority of funding agencies). Unfortunately, the word "monitoring" is sometimes still perceived negatively in scientific panels as synonymous with exploratory research with no clear hypothesis (Pereira et al. 2010; Lovett et al. 2007; Nichols and Williams 2006). Even when there is an intention to take basic ecological monitoring data (and the many recent demonstrations of

BOX 6.2

## The Rich Get Richer in Invasion Ecology

**TOM STOHLGREN**

Influential ecologists such as Charles Darwin and Charles Elton believed that habitats low in native-species richness were more prone to invasion by alien species. Small-scale experiments in artificially constructed "communities," with 1-$m^2$ plots in protected old fields, seemed to agree with earlier observations (Tilman 1999). Based on these carefully controlled experiments, scientists concluded that ". . . diverse communities will probably require minimal maintenance and monitoring because they are generally effective at excluding undesirable invaders" (Kennedy et al. 2002).

Meanwhile, my field crew and I wanted to measure patterns of native and alien plant-species richness in natural communities that include a montane meadow and forest in Rocky Mountain National Park, Colorado. We were also interested in how the scale of observations influenced our results, so we established a series of multi-scale vegetation plots. Each 1000-$m^2$ plot contained one nested 100-$m^2$ subplot, two 10-$m^2$ subplots, and ten 1-$m^2$ sub-plots. Preliminary data suggested that at most spatial scales (10-$m^2$ to 1000-$m^2$), there was a positive relationship between native-species richness and alien-species richness. Results were highly variable for the 1-$m^2$ subplots. We immediately replicated the sampling in the Central Grasslands, with additional sampling in four states. We found the same general patterns and reported our findings (Stohlgren et al. 1999). Additional soil sampling confirmed that habitats high in soil nutrients and soil moisture generally supported more native and alien species than habitats low in light, water, and nutrients. Our hypothesis was that environmental conditions that fostered native species also fostered alien species—the rich would get richer (Stohlgren, Barnett, and Kartesz 2003). We tested this hypothesis by comparing nutrient-rich riparian zones with adjacent, drier upland sites. Again, we found the same predictable results (Stohlgren et al. 1998). We then gathered independent data sets (data we didn't collect) from other observational studies, including herbaria records and forest health monitoring plots from across the conterminous United States. We were alarmed to discover that the "rich get richer" pattern of invasion is a widespread pattern at most spatial scales. Further observational studies in natural ecosystems cast serious doubt on the importance of competition and biotic resistance in containing invasions (Stohlgren et al. 2008). We conclude the exact opposite of Kennedy et al. (2002): "Diverse *natural* communities will probably require *maximum* maintenance and monitoring because they are generally *ineffective* at excluding undesirable invaders."

the value of monitoring is starting to change perceptions of this activity), many government agencies, many countries, and even whole regions of the world lack the resources needed to collect ecological data in a systematic and sustained manner. And once the data have been taken, they may be inaccessible because of the digital media problems we discussed in Chapter 4, or because they haven't been carefully curated.

In the early days of ecology, the way to lock in observations was to collect organisms and save them in natural history museums. The ability to shoot, stuff, fix, and preserve biological specimens was considered elemental and indispensable for a scientist, who could make his mark by the size of the collections attributed to him in prestigious museums. Both the teaching of these collecting skills and the central role of natural history museums faded throughout the twentieth century, but it is now being rediscovered and enhanced in the genetic age (Wandeler, Hoeck, and Keller 2007; Graham et al. 2004) when whole new discoveries about the relationships of organisms and their historical population sizes, among other phenomena, can be assessed by using preserved genetic material. For example, genetic samples stored in museums can help us understand population demography and genetics in historical times. Craig Miller and Lisette Waits (2003), analyzing 110 museum samples of Yellowstone grizzly bears collected in the twentieth century, found that genetic diversity and the size of these populations have been low historically, so that their current small population size may not reduce genetic diversity as quickly as expected. Even the low-tech simplicity of a museum record, which plainly states that "this organism lived in this area at this time," can be surprisingly useful for studies of how species may change over time in their distribution, demography, and genetics, especially given the uncertainties of climate change (Johnson et al. 2011; Lister 2011).

As samples for herbariums and museums were often collected as a matter of course during an expedition, with no particular question in mind to limit the scope of the collection, impressive amounts of information are available in these institutions. Of course, biases can arise due to both the non-quantitative nature of most collections and the fact that collections are not sampled at random from a larger population. We can't know, unless detailed descriptions were included with a collection, how much effort the original researcher put into finding and collecting the individuals in a given museum lot. There may also be an overrepresentation of certain species that were especially attractive to naturalists of a given time, and even experts on a certain group (gastropod snails, for example) may focus disproportionately on certain species within that group.

These biases can become problematic when they leave collections disproportionately underrepresented in species that indicate ecological phenomena of interest (Bebber et al. 2010). For example, a bias among early botanists against collecting introduced plants has reduced the opportunities to gain a clear understanding of the history and spread of non-native plants. This is especially relevant in countries that have only recently developed a greater critical mass of botanists. In Chile, for instance, this collection bias against non-native species and ornamental conspicuous species has limited the number of samples of species such as *Pinus spp.* and other very invasive trees and shrubs. As a result, ranges and resident times for these species can be highly underestimated by using herbarium records alone. Collections should be planned to maximize relevant data collection in the field, given limitations in funding and human resources (Bebber et al. 2010).

There are also ethical issues behind all those pinned lepidoterans and skinned mammals. Some people oppose the idea of collecting species just for the sake of it, and this opposition group understandably grows when we are dealing with rare or endangered species. But the ethics aren't black and white. A recent documentary film, *Ghost Bird*, which addresses the question of whether the ivory-billed woodpecker is really extinct, poignantly showed the paradox of collecting for museums. Museum-preserved skins of ivory-billed woodpeckers collected by naturalists provide a critical link enabling today's researchers to check if their fleeting observations in the swamps of the southeastern United States are really remnant populations of living birds. But those collections themselves, made in the early twentieth century, likely had a devastating effect on the then-dwindling ivory-billed population.

Curators have been grappling with these questions and trying to answer them in a modern context. Digital photography offers the opportunity to inexpensively archive many details of organisms that are then left to go on living in the field. Genetic data can be collected and archived based on small, non-lethal tissue samples, even in the case of endangered species. For example, whales can be sampled both by photographing their distinct markings and by obtaining small samples for genetics research by lancing the blubber when they surface (provided that the researchers have waded through a mass of federal and international regulation to get the appropriate permits).

Obtaining the funding for museum collections is always difficult and becomes more so as costs increase due to the accumulation of collections (Mazzarello 2011). Museums can store data but they can't promise quick or definitive answers to ecological questions. For funding agencies and foundations looking for a sure bet in a short-time frame, the pace and primary mission of museum activities appears relatively less attractive than well-constrained experiments that can state clearly what they will yield. Large and small museums alike are facing cutbacks, and taxonomists and curators are becoming extinct (Pearson, Hamilton, and Erwin 2011). Although funding is critical in every country, the problem is acute in developing countries where the push to become scientifically productive has channeled most funding into initiatives with high return in terms of indexed papers, leaving museums and herbaria practically unfunded.

There are strong parallels between museum collections and long-term ecological data taken in the field, in terms of practical limitations, potential biases, and potential benefits. One difference, however, is that while museum collecting began as a basic science method for cataloging the diversity of life, long-term monitoring developed with a more applied goal in mind.

Foresters and range managers started long-term ecological plots with the practical goal of maximizing yield, and their early attempts have served as the basis for long-term ecological data collection. Foresters were interested in understanding the dynamics of biomass in forest stands with and without silvicultural treatments. Thus they introduced the concept of both observing changes in natural systems and also manipulating ecological conditions. Trial plots and monitoring plots were quickly established across the globe, especially in countries in Europe and North America having a forestry tradition (Tomppo 2009). By looking mainly at biomass variables (e.g., diameter, height, species composition) oriented to increase forest productivity, foresters started to accumulate the information needed to understand the ecology of these forests. Similarly, range managers interested in the effect of grazing on the productivity of pastures set up exclosure plots that kept livestock away, and these yielded early evidence for concepts such as carrying capacity, top-down control, and ecological succession (Johnson and Matchett 2001). A good example of grazing plots that have been monitored for many years occurs in Yellowstone National Park, where long-term plots have served to help ecologists understand the dynamics of herbivory, top predators, and nutrient cycling, as well as conservation issues such as the effect of reintroducing wolves (Verchot, Groffman, and Frank 2002). Unfortunately, just as museum records give only a narrow (and sometimes biased) window into larger and more complex data, field plots that were not intended to facilitate the understanding of a whole ecosystem can be limited in their utility today, and their results should not be scaled up without further considerations of other factors that operate over larger scales (see, for example, Freilich et al. 2003).

All long-term monitoring and sampling programs, especially those in the developing world, went through periods of neglect in the twentieth century (Southward, Hawkins, and Burrows 1995), and they continue to struggle for continuous support today, not unlike the museum collections discussed above. In some cases, long-term records stored through natural processes, as well as historical documents, may help close the data gaps in time and among regions of the world. For example, to study the influx of invasive species in central Chile, pollen records in lakes surrounded by heavy agricultural and land development may contain important information about community composition. These records may be complemented with historical diaries, early botanical records, and image archives, all of which inform us about the species that were present in the landscape at that time. The combined use of all these observational tools may help to uncover the dynamics of native and non-native species in the early Spanish settlement period (1500s–1700s), in which information is very scarce due to the lack of local botanists in this era.

### Networks of Observers and Observations

Some of the practical and methodological problems with long-term data collection can be ameliorated by relying on networks. The global characteristic of ecological questions, the pressing need for rapid answers, and the globalization of human society have made international scientific networks one of the most powerful tools for research in the twenty-first century. What do we gain by networking? Obviously, the extent and depth of ecology has grown exponentially in recent decades. We cannot expect that a single ecologist will be able to handle all the information and methods available for answering a broad question in ecology. Networks allow people to exchange ideas, information, and methodological tools at a speed unprecedented in the history of ecology. They also allow researchers to test their hypotheses over large spatial and temporal scales by using data collected at multiple sites over variable periods of times. In other words, a question that could only have been tested in one single location in the past, with very little opportunity to scale up, can now be tested globally. We think that the most important networks also allow people from different cultural, educational, and scientific backgrounds to discuss hypothesis and ideas, thus breaking down institutional and intellectual barriers in ecology.

Under the Long Term Ecological Networks (LTEN) scheme conceived in recent decades, broad ecological questions are developed along with a network of sites to collect data that will help us address these questions. The power of an LTEN is in having continuous data, taken in a similar manner, over a large temporal and spatial scale. There are strong regional differences in the LTEN of the world. For example, the United States has both a network of long-term ecological research (LTER) sites and a National Ecological Observatory Network (NEON), which spent $80 million in planning and has nearly a half-billion-dollar operational budget, with aims to deploy advanced sensors and observations across a network of fixed "core" sites and flexible "relocatable" sites (Tollefson 2011). Yet LTEN have only slowly percolated to all regions, countries, and ecosystems, especially in less developed countries where some of the most important reservoirs of biodiversity are located. Groups such as International Long Term Ecological Research (ILTER) are trying to deal with these gaps by providing a "network of networks" linking the activities of monitoring networks throughout the world. Nonetheless, networks are still lacking in central and northern Africa, the Middle East, and much of Asia (www.ilternet.edu).

While networking can be applied to any single issue in ecology, some ecosystems, such as mountains, are particularly amenable to the network approach. Mountains are essentially globally replicated steep elevational gradients where we expect to see rapid transitions in ecological phenomena; they can also be used as spatial stand-ins or proxies for temporal changes expected with climate warming (Pauchard et al. 2009). In 2005, a group of scientists started a network to understand how mountains were being affected by plant invasions. The Mountain Invasion Research Network (MIREN) now encompasses people from 11 regions of the world (Figure 6.1). By using a common and very basic sampling protocol designed to use a hierarchical multi-scale approach, MIREN has illuminated patterns of non-native plant distributions globally. For example, Seipel et al. (2011) found that the hump-shaped or linear decline of non-native species richness with increasing elevation is a repeated pattern in all seven regions sampled, independent of the absolute elevational range studied. Furthermore, the ranges of non-native species reflect a very generalist behavior of these species, while native species show a much more specialized niche across the elevational range (Alexander et al. 2011). Using literature and available data sets, MIREN has found that non-native species in mountains are more similar to their lowland counterparts than to those of other mountains (McDougall et al. 2011). More refined observational techniques can also be used, for example to test whether a species grows differently because of its genetics or environmental differences among mountain regions. Experimental approaches can also be integrated into this type of network. For example, climate change (e.g., warming) experiments

Figure 6.1

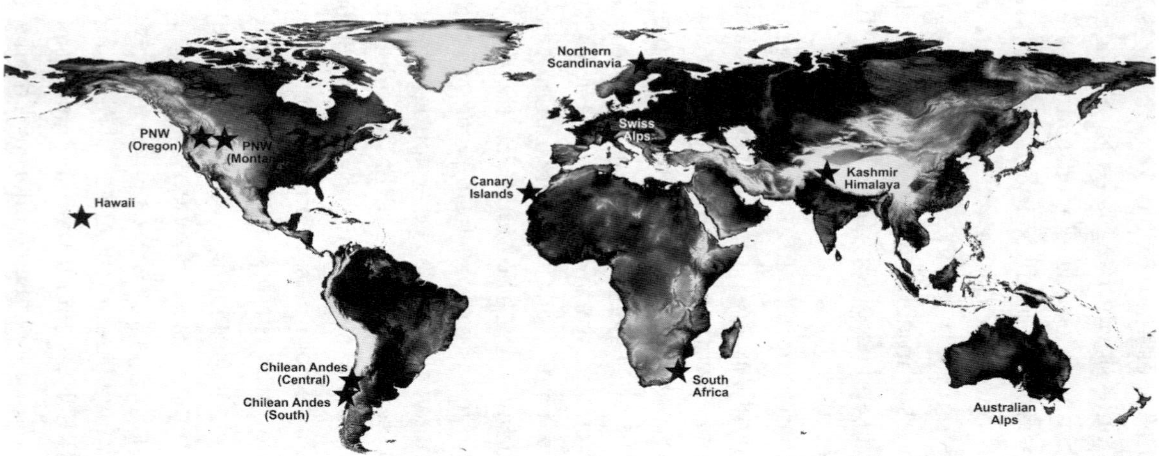

*The Mountain Invasion Research Network (MIREN) encompasses 11 regions across the globe. It is designed to allow for global comparisons using a hierarchical approach, which allows for more-specific studies within climatically (e.g., tropical, temperate) or continental (e.g., South America, Europe) clusters of regions (Pauchard et al. 2009).*

or translocation experiments may help to test whether native and non-native species will be able to withstand rapid increases in temperature and whether they will be able to disperse upward on mountain hills (e.g., Alexander et al. 2011; Poll et al. 2009).

### Challenges and Opportunities in Analyzing Observational Data

Recording, collecting, and storing large amounts of data, as is being done in these networks, has much more value when the data are systematically analyzed for their ecologically relevant patterns and relationships. Sometimes making sense of data is just a matter of running simple descriptive statistics (means, variances, etc.) or simple correlations (e.g., linear regressions) but more often large, complex data sets are too nuanced to be described by a simple statistic. As K. Robert Clarke and R. M. Warwick state in the beginning of their book on detecting ecological changes, such statistics are "technically feasible, though rarely very informative in practice, given the over-condensed nature of the information utilized" (Clarke and Warwick 2001). As networks help close intra-planetary data and knowledge gaps, they will begin to generate a flood of data even as the still-neglected regions and questions face a data drought. Ideally, data analysis can cover the range from data-poor to data-rich situations, but to do so these tools must be flexible and adaptable.

Fortunately, one of the benefits of practicing ecology now is the great flexibility with which we can analyze data. This arises both because there are so many tools that are easily accessible and because the tools themselves are more flexible than the "parametric," usually univariate, statistics that dominated late-twentieth-century ecology. So-called parametric statistics, such as the ubiquitous analysis of variance (ANOVA)—which will always have an important role in detecting ecological change—nonetheless carry the baggage of several underlying assumptions that must be met for parametric statistics to provide accurate results. Chief among these assumptions is that data are pulled from a normal distribution, or something that can be mathematically transformed (for example, by taking the *log* of all the data) into something that closely approximates a normal distribution. This is a problem for many of the data we'd like to look at in these days of rapid change, heavy anthropogenic impact, and high unpredictability. Nassim Taleb, a former stock trader writing in his book about unpredictability, *The Black Swan* (2007), shreds our assumptions about the prevalence of normal curves in the real world. While he concedes that some things marked by physical properties with constrained limits, like heights of people in a population, are relatively predictable and tend to fall out in a normal curve, he argues that, nevertheless, the really interesting stuff, the complex and unpredictable phenomena, are almost never defined by a normal curve.

Not surprisingly, much of the ecological world falls into the "non-normal" category, but even the things we'd expect to be normal, like the sizes of individual snails in a population, don't turn out that way in an age of massive human impact on ecological systems. Most species' size distributions for example, which may have been normal at some point, are now completely cut off at the right end of the curve (where you'd find all the large individuals if sizes were laid out on a graph). If the species are highly managed with minimum allowed sizes for harvesting, you'll find size distributions that fall off dramatically right after the minimum size limit.

Non-normal curves in nature are also popping up in distributions that our intuition wants us to believe "should" be normally distributed, even in those distributions we've been told are perfectly normal. Part of Rafe's thesis work showed that species' distributions of abundance across their ranges don't follow a normal-shaped pattern with highest abundances at the center of the range (Sagarin and Gaines 2002a), as was assumed throughout the twentieth century (Sagarin and Gaines 2002b). This long-standing belief has now tumbled in just a few years as dozens of investigators have gone out and simply observed species' population distributions across their ranges and have failed to find normal curves (Sagarin, Gaines, and Gaylord 2006).

But rather than being debilitating, abandoning the assumption of normality can be tremendously liberating and empowering. We are faced with a whole new world full of very weird distributions—species that are very abundant and then suddenly disappear, populations without any mature females, and physical conditions that switch from benign to catastrophic in a relative instant—to name a few non-normally distributed phenomena. Fortunately, we have an abundance of tools to guide us through this strange, uncertain world. Many of these tools fall into the realm of "multivariate" statistics, which can handle large numbers of individual species (or species assemblages or genes) sampled in different environments, and each may be associated with multiple attributes (e.g., size, concentrations of toxins, parasite load) that themselves aren't normally distributed. Typically, multivariate stats make some kind of similarity (or dissimilarity) matrix of all the data, which calculates how much alike (or different) each possible pair of data in a data set are to one another and ranks them accordingly. The relationships in this matrix can then be visualized as a dendrogram (a "tree" of relationships, often used to compare genetic data, with close "branches" representing tight similarity between samples) or as clusters of similar samples spread across a space.

An advantage of these approaches is that, although they can be computationally complex, they are conceptually straightforward (Clarke and Warwick 2001). In their highly readable guide to the complex-sounding multivariate technique called "non-metric multidimensional scaling," Clarke and Warwick illustrate how these approaches work by creating a rank-ordered similarity matrix of 24 towns in the United Kingdom based on the distance between each pair of sites. The resulting outcome in the program is a field of data that looks just like a map of the UK. In more ecological data, multivariate statistics can show, for example, whether an assemblage of species changes before and after a disturbance, or if communities closer to a sewage outfall are significantly different from communities living in cleaner water.

Each facet of any given multivariate analysis can also likely be broken down and analyzed. Spatial statistics help ecologists quantify relationships between observations in space and can uncover hidden patterns, such as the "autocorrelation function" (a measure of the distances at which relationships between sets of sites can be discerned), that may point to underlying ecological mechanisms. Time series analysis similarly looks for temporal relationships, such as recurring cycles or lags between a causal agent and its effect. Bayesian methods, which ask whether data fit a set of probabilistic assumptions, have been introduced to ecologists (Clark 2005; Ellison 2004) and are being used to help filter large data sets, to directly analyze basic and applied ecological questions (Wade 2000), and even to ask whether animals themselves use Bayesian inference in their behaviors (Valone 2006). Bayesian analysis seems particularly well suited to the approaches we discuss in this book because it mimics the way keen observers face complex ecologies by developing some assumed probabilities about what is being observed based on past knowledge (these are called "prior probabilities" in Bayesian terms) and then testing to see how well these prior assumptions fit.

As we stated at the outset of this chapter, it's not that these techniques are all brand new—some have been used by ecologists for decades—it's just that the diversity of techniques seems to be exploding as the problems we encounter are more complex and multidisciplinary and they become more computationally accessible with increased computing power and networked sharing of techniques. Indeed, there are specialized commercial software programs that can run each of these types of analyses, but the open-source statistical language "R" provides these analysis packages for free and is also continually updated with new statistical tools submitted by a large community of users. The updated list of R packages for environmental analysis (cran.r-project.org/web/views/Environmetrics.html) contains dozens of such programs that can be added to the already powerful basic R installation.

Not all of the tools used to understand complex ecological systems require complex mathematical and computational tools. Synthetic approaches, such as meta-analysis, which formally and statistically analyzes multiple studies to generalize concepts from many individual small-scale experimental manipulations or individual observational investigations (see Harrison 2010, for a good introduction), are a way to build large-scale understanding out of widely dispersed data sets. A meta-analysis does more than simple "vote counting" of which studies did or did not show a significant result; it also weighs the importance of studies differently based on the sample size and design of the study and the congruence of results between studies. Meta-analysis was used in the first two comprehensive reviews of twentieth-century climate-change effects on natural systems (Parmesan and Yohe 2003; Root et al. 2003), which primarily used observational studies. By contrast, Bradley Cardinale and colleagues (2006) used meta-analysis of many experimental studies across trophic groups (e.g., herbivores, carnivores, detritivores) and habitats to show that declining species richness can affect ecosystem functioning (such as how resources are depleted), but that the exact pattern of change can't be predicted in theory. The value of this approach is exemplified by the success of the National Center for Ecological Analysis and Synthesis (NCEAS) in Santa Barbara, California, which was set up as a think tank for multi-investigator, multidisciplinary projects that attempt to derive new insight through combining and analyzing existing data sets and revising existing conceptual constructs. Although founded only in 1995, NCEAS quickly rose to the top 1 percent of over 39,000 ecological institutions based on impact factor (Hackett et al. 2008) and has become a model for at least 17 new ecological institutions internationally (S. Hampton, pers. comm. to RDS).

Finally, a lot of the data that ecologists are getting now comes in the form of words, not numbers. These include qualitative data, but also semantic data that can be quantified. For example, one of Rafe's students, Mary Turnipseed, used a popular text-analysis program called NVivo to scan and analyze thousands of public comments on U.S. President Obama's proposed National Oceans Policy for common attributes of comments that mentioned "public trust doctrine" (the subject of her thesis work). Combining qualitative and quantitative approaches that use numeric and other types of data falls into the general category of "mixed methods," which has been gaining traction in ecology, as well as in public health, education, and financial analysis, all areas where complex realities produce complex data sets.

Because mixed methods invariably combine fields of knowledge and approaches that have been treated separately for a long time, a lot of the literature on mixed methods gets into epistemology—the philosophy of how we know what we think we know. Epistemological questions can take us down a rabbit hole into deep warrens of discussions, but they are not completely esoteric. In fact, the way in which philosophical questions about what is the nature of science have been answered has held back a full appreciation of the validity of observational approaches, as we discuss in the next chapter.

CHAPTER 7

# Is Observation-Based Ecology Scientific?

Questions and criticisms that arise around observational approaches boil down to one fundamental question: "Is this really science?" This question could be asked of any kind of research, but because observational approaches have been out of the scientific mainstream for a long time, and because they invite so many nonscientists as well as investigators from the so-called soft sciences to be part of the life sciences, it is frequently directed at observational studies. So now that these kinds of studies are being integrated into science, the natural follow-up to the fundamental question becomes "Are observational approaches scientific?"

There is not just one answer to the question "What is science?" and this means that a number of different philosophies of science have arisen over the years, some more influential than others. Just as the techniques and goals of ecological science are changing, so too will its underlying philosophies. In this chapter, we briefly discuss the most influential scientific philosophies of the last century, particularly falsification and "strong inference." These philosophies complement the experimental approach to ecology well, but they are sometimes difficult to reconcile with observational approaches. It is because of this mismatch that some have argued that observational approaches are not scientific.

We live in a very different ecological world from even a few decades ago and observational approaches are more powerful than ever before.

We suggest that we live in an era when the underlying philosophy of science can expand to be more inclusive of observation-based studies while still justifying experimental and theoretical methods as well. Fortunately for us (because neither of us is a philosopher), other philosophers of science, both past and present, have made this realization and have put forth several variations of more-inclusive ecological philosophies, which we adopt here.

Early ecology proceeded by building the observational work of naturalists into theories that were tested against further observations. The philosophical match to this approach is found in early-twentieth century ideas of the "logical positivists," who believed that scientific theories could be tested by verifying them against observable phenomena. This was an inductive approach that built understanding from layering different levels of evidence until a theory was confirmed.

But the mid-twentieth-century science philosopher Karl Popper felt that inductive reasoning could too easily fall victim to pseudoscience. He espoused a philosophy that drew a clear and impenetrable line between what was science and what wasn't science. In contrast to the logical positivists, Popper argued that science is defined not by what can be proven or verified, but by statements that can be disproven, or falsified. Therefore, a conclusion such as "the Earth was created by a giant invisible octopus" is not scientific because no other explanation was falsified to arrive at this conclusion. Moreover, it is difficult to conceive of any test that has the ability to disprove this kind of conclusion. However, you can use science to test (and possibly falsify) a conclusion such as "the Earth is 6,000 years old," because we now have various methods for accurately dating rock strata and fossils and they have repeatedly yielded results indicating that Earth history must be measured in billions of years, not thousands. In this case, the conclusion would be updated by a scientist as "the Earth is billions of years old," a statement that both reflects our current understanding, yet is still falsifiable. Science is the process of making a statement and having a means to falsify that statement, according to a Popperian view.

Popper's philosophy was timely. As we discussed in Chapter 2, the rise of molecular biology, which provided precise, repeatable conclusions, put pressure on ecologists to develop more-rigorous approaches to achieving ecological understanding. Popper's standard of falsification offered one way to ensure that ecological studies reached the same idealized scientific standard as physics and molecular biology. But in order to be widely accepted in the ecological community of practice it needed to be more than just a philosophical argument (Buck 1975).

### The Strengths and Weaknesses of Strong Inference

Against this background, John Platt's "Strong Inference" paper in *Science* (Platt 1964) provided a practical mechanism for applying Popperian falsification to biological sciences. The strong inference approach involves developing multiple falsifiable hypotheses that can be tested, pairwise, against one another until one most likely explanation remains at the end, and such tests can be repeated by different investigators to ensure the validity of the result. Platt gives an all-or-nothing quality to strong inference, suggesting that it is "*the* method of science and always has been" (Platt 1964, p. 347). Critically, in terms of the paper's influence, Platt pitched his approach in tantalizing terms, suggesting that strong inference was the pathway to make all of biology more like molecular biology: "I believe we may see the molecular-biology phenomenon repeated over and over again, with order-of-magnitude increases in the rate of scientific understanding in almost every field" (Platt 1964, p. 352).

There is no doubt that "Strong Inference" touched a nerve. The article has been cited over 1,000 times by other scientific articles in ecology and in fields well beyond ecology, such as psychology, medicine, and economics. Many of these papers cite "Strong Inference" to justify the research approach they use. Others use the paper to rally their colleagues to adopt the strong inference approach in their own field. The strong inference approach is also reinforced from the most elementary level of science students all the way to practicing scientists. When the "scientific method" is taught to students, it usually takes the form of a strong inference approach. Papers by prominent ecologists have called approaches that don't strictly subscribe to the strong inference method "simple" and "seductive" (Simberloff 1983), even "weak" and "soft" (Elner and Vadas 1990). Proposals to the National Science Foundation (NSF), even those to support long-term research, are implicitly expected to follow a strong inference approach where the hypotheses, and expected tests of those hypotheses, are stated up front.

This leads to a paradox and to perverse incentives among practicing ecologists. We need funding to go out and find interesting things in the environment at the large scales in which environmental change is happening, but we need to create hypotheses to justify the funding even though the truly interesting hypotheses can't emerge until after we've looked at a system broadly for patterns that warrant further investigation. As a result, we are forced to state hypotheses that we know to be trivial, or else we risk grant rejection under the criticism that the proposed work is "merely exploratory" or a "fishing expedition." As Jacob Weiner has asked, "How many ecologists write their grant proposals and papers in terms of hypotheses and tests retrospectively, even though the work was not conceived in this way?" (Weiner 1995).

This is potentially more of a problem than simply "going along to get along"; psychologists have found that scientists who state predetermined hypotheses tend to "satisfice"—that is, make subtle changes to their data interpretations to better fit their predetermined hypotheses (Garst et al. 2002). Even before we had psychological tests and jargon to name this phenomena, early ecologists like Ed Ricketts recognized the potential for bias to arise when an investigator started with a hypothesis about "why" something exists, rather than starting with observations of it. He argued that this led to a "wish fulfillment delusion," noting,

> When a person asks "Why?" in anything, he usually deeply expects, and in any case receives, only a relational answer in place of the definitive "Because" which he thinks he wants. But he customarily accepts the actually relational answer as a definitive "because." (Ricketts 2006)

In other words, even if we get only a partial answer, we assume it is a complete answer if it tends to match our preconceived notions.

In this context, we can mechanistically dissect on our initial concerns about bias that we raised in Chapter 3. Here we suggest that biases that have been deeply embedded in scientists through cognitive processes and through their experiences and cultural backgrounds may be amplified in a framework where only limited options for further exploration are provided. In *Field Notes on Science and Nature,* a beautiful new book on the science and art of field notes, contributor George Schaller argues that even the act of observing can be biased if we try to constrain it into preconceived categories. He reports that he is careful not to rely solely on pre-made checklists for his field notes because "an important detail may be ignored or considered irrelevant and discarded because it lacks a discrete category on the list. It is often an anecdotal event that offers special insight" (Schaller 2011). The argument of Schaller and Ricketts and the psychologists writing about "satisficing" isn't about eliminating all bias, but rather that predetermining what you are looking for (those data that fit or do not fit a particular hypothesis) is likely to increase a tendency toward whatever the observer is already biased toward.

A number of other practical, statistical, and logical flaws to the strong inference approach have been well documented (Lawson 2010; Holling and Allen 2002; Pigliucci 2002; Dayton and Sala 2001; Beyers 1998; Weiner 1995; Francis and Hare 1994; Wenner 1989; Quinn and Dunham 1983). Strong inference relies on testing a continually dividing tree of binary hypotheses, but many ecological phenomena occur across a continuum (O'Donohue and Buchanan 2001). Ecological questions often can't be reduced to "reject" or "do not reject" a hypothesis. For example, in toxicology studies the question is almost never "Is this substance toxic or not?" but "How much of this substance is toxic, and under what conditions?" Moreover, as we've argued throughout this book, many ecological phenomena of interest cannot be manipulated in a framework amenable to strong inference. They often cannot be replicated or randomly assigned to treatments in the manner of a well-crafted experimental hypothesis (Hilborn and Ludwig 1993). Conducting multiple tests of multiple hypotheses of interest isn't possible when the question is, for example, "How will the Fukushima nuclear disaster affect coastal ecological communities in Japan?" In many cases the kind of randomized approach favored by statisticians to alleviate potential bias in planned experiments is impossible to achieve. As Ray Hilborn and Donald Ludwig wryly noted of the ecological effects of the *Exxon Valdez* oil spill, "Oil did not strike at random" (Hilborn and Ludwig 1993, p. 551). Ultimately, some of the most important ecological questions can't be answered with data that can be forced into a strong inference approach.

Another problem with applying the strong inference approach "formally and explicitly and regularly," as Platt urges us to do, is that many of the most important discoveries in science occurred serendipitously, as unexpected findings in a data set or from accidents in a procedure (O'Donohue 2001) or even from physical accidents resulting in near-death experiences, as Ricardo Rozzi recounts in Box 7.1. Scientists will often make testable hypotheses out of these findings—for example, when a large decline in zooplankton was found unexpectedly in the California Cooperative Oceanic Fisheries Investigation (CalCOFI) data set (which was designed to track sardine populations), other scientists then hypothesized that they would find concomitant declines in seabird populations (Ainley et al. 1995). Finally, William O'Donohue and Jeffrey Buchanan also suggest that strong inference is a problematic set of guidelines for science precisely because it is a set of guidelines. They point out that some of the most important advances in science, such as the Copernican revolution, occurred because scientists broke the accepted rules of the day about how to do science. The observational approaches we discuss here aren't exactly the work of rule-breaking revolutionaries, but rather of rule-broadening pioneers. They don't require us to reject any approaches to science, but rather to expand those methods that, under the right circumstances, may be considered scientific.

Critiques of strong inference haven't necessarily taken hold in the mainstream of science, in part because Popper and Platt's ideas of what science "should be" actually fit perfectly well with an experimental approach. Almost by definition, a well-designed experiment tests multiple predetermined and falsifiable hypotheses in a stepwise fashion. Many ecologists who have developed widely accepted paradigms based on experimental approaches may consider it axiomatic that experimen-

BOX 7.1

## Changing Lenses to Observe, Conserve, and Co-Inhabit with Biodiversity: Serendipity at the Southern End of the Americas

RICARDO ROZZI

In March 2000, I embarked on an expedition to the Cape Horn Islands at the southern end of the Americas, guiding a group of bryologists led by Bernard Goffinet in the search of Splachaneceae or "dung" mosses that grow on the bones of whales beached at the margins of peatlands and bogs. After surviving several storms while navigating in a tiny fishing boat, we frantically initiated the search for the mosses on Navarino Island. While jumping over the bogs, I became separated from the group and fell into a peat bog. I started to sink, sure that this would be a quiet, natural death. While sinking, I observed the astonishing diversity of mosses around the pond, and I thought, "If I am a biologist and do not have knowledge of this diversity of plants, what about the decision makers and teachers in Chile?" Some years earlier, I had participated in committees charged with identifying priority sites for conservation in Chile and Latin America, which were based only on vertebrates and vascular plants. In accordance with that framework, the Magellanic sub-Antarctic *ecoregión* was classified as unknown or of low priority for conservation.

Fortunately, Bernard and the team found me in the swamp after a couple of hours, just before I completely disappeared. I survived the episode, but the image of the exuberant diversity of mosses became fixed in my mind. I began a systematic bibliographic review of the bryophytes in Chile and complemented those results with floristic inventories initiated with Bernard, William Buck, and other bryologists in Cape Horn, and *eureka*: we discovered that the Magellanic sub-Antarctic *ecoregión* constitutes a world hotspot of mosses and liverworts diversity!

In less than 0.01 percent of the planet's terrestrial surface we find more than 5 percent of the bryophyte species known to science. In the austral *ecoregión* mosses and liverworts are more speciose than vascular plants, contrasting with the ratios of vascular/nonvascular plants found in lower latitudes (Rozzi et al. 2008). This discovery stimulated us to propose a "change of lenses" to observe biodiversity: to assess high-latitude floristic diversity, we should not base inventories merely on vascular plants, but also include the nonvascular ones. Rather than a narrow set of global indicator groups, ecoregional- or biome-specific indicator groups are needed for effective assessments of biodiversity.

The "change of lenses" had implications not only for observing biodiversity ▸

▸ but also for conserving it. The high diversity of sub-Antarctic bryophytes was one of the central arguments for the creation of the UNESCO Cape Horn Biosphere Reserve in 2005. This largest biosphere reserve in southern South America was created based on organisms that, up until now, had rarely been perceived and valued in the region, the country, and the worldwide conservation community.

Finally, the "change of lenses" to observe and conserve biodiversity led us to an ethical change to co-inhabit with the sub-Antarctic biodiversity. Together with children at the local school in Cape Horn, we composed the metaphor: "miniature forests of Cape Horn," through which mosses, liverworts, lichens, insects, and other organisms were perceived as *co-inhabitants* rather than mere "natural resources." Children observed the mosses' reproduction, growth, and ecological interactions while lying and breathing close to them. Through these observations they cultivated a sentiment of empathy realizing, in their words, that "we—humans—also breathe, reproduce, grow, and interact with other organisms." Through observation and direct "face-to-face" encounters with mosses in their native habitats, children (and researchers) understood both the intrinsic and instrumental values of mosses; the latter based on gaining an understanding about the role that mosses play in the regulation of water flow and quality in the sub-Antarctic watersheds.

These field experiences, in turn, stimulated the invention of "ecotourism with a hand lens," an activity triggered by the appreciation of the beauty, diversity, and socio-ecological relevance of this little flora that usually remains under-perceived by citizens, teachers, and decision makers. Ecotourism with a hand lens has attracted growing interest from tourists, who are arriving in the area in rapidly rising numbers. In collaboration with the children, graduate students from the Universidad de Magallanes, Francisca Massardo, and other researchers, as well as regional authorities, teachers, artists, engineers, architects, and other professionals, we decided to create the "Garden of Miniature Forests of Cape Horn" to implement the novel ecotourism activity, and to promote the conservation of the sub-Antarctic bryoflora at the Omora Ethnobotanical Park. The building of the garden and interpretive trails helped to show that to conserve and learn sustainable forms of co-inhabitation it is not enough to change our conceptual lenses, but it is also necessary to implement areas for conservation and to conduct practices of observation in the field.

In this way, a transformative field experience of observation triggered a sequence of changing lenses to (1) assess, (2) conserve, and (3) cultivate an environmental ethic of co-inhabitation with biodiversity.

tal approaches are necessary in order to understand ecological systems. Robert Paine, whose extensive experimental work on the intertidal zone of Tatoosh Island, Washington, in the northwestern United States laid the groundwork for the widespread concept of "keystone predation," recently argued, "Whatever the reasons for the success of microecology, the evidence clearly indicates that continued attention to small-scale experiments and functional roles holds the greatest promise for managing our world for a sustainable future" (Paine 2010). Daniel Simberloff, who conducted some of the classic experimental work on island biogeography theory, considered the truly valuable ecological work to be that which contains "an unequivocally falsifiable hypothesis and a system sufficiently simplified, by whatever means, to allow an unambiguous test of the hypothesis" (Simberloff 1983).

### Why Observational Science Isn't Considered Science

Even when the observational approach is acknowledged, it is often assigned second-rate status. In a paper by 16 ecologists resulting from a National Science Foundation–sponsored workshop to determine the future funding priorities for population and community ecology, the authors noted that "although experimental approaches will always be required to demonstrate mechanisms underlying ecological phenomena, observational studies complement and expand on what can reasonably be studied in an experimental context" (Agrawal et al. 2007). In other words, there is a sense that experiments should take primacy in ecological study (to be "complemented" by observation). Additionally, there is a tightly held belief that only experiments can unearth underlying ecological mechanisms (Paine 2010; Simberloff 2004).

The belief in the ability of experiments alone to uncover mechanisms in turn is built upon four common criticisms of observational approaches, which argue that: (1) observation-based studies find patterns, but patterns can't be used to infer process; (2) observation-based studies rely on the flawed approach of induction, rather than the more precise deductive approach, to reach conclusions; (3) they are just a collection of unreplicated anecdotes; and (4) they rely too much on correlations between variables. Here we address these criticisms in turn and then return to the question of developing a more inclusive framework for reaching ecological understanding in which observational, experimental, and theoretical approaches can work harmoniously, creating synergism among them.

First, it is true that most observational studies collect, or borrow, lots of data and look for patterns in the data. Weiner and many other ecologists, in fact, consider pattern finding to be the primary role of an ecologist (Weiner 1995), but there is also a strongly held belief among other ecologists that such pattern finding must be accompanied by manipulations to get at the underlying mechanisms. There are good reasons to be cautious about inferring a process from a pattern—any observed pattern could be the result of multiple different causal factors (McIntire and Fajardo 2009). But making sure that you are not ascribing causation to the wrong factor is really a challenge common to all forms of ecological, and scientific, understanding.

Given enough data and the right analytic methods, it is possible to link causal factors to patterns and even to test hypotheses based on pattern data alone. Eliot McIntire and Alex Fajardo recently argued that "the connection between space and process is in a period of rebuilding after being rejected by numerous authors over the past 50 years" (McIntire and Fajardo 2009). The authors suggested that many of the critiques of pattern analysis are based on incomplete or outdated analyses that can now be improved with new statistical models that are, in turn, made available by better computing technology. McIntire and Fajardo recommend a method of hypothesis testing using spatial patterns as a surrogate for time or other ecological factors that are difficult to manipulate. In their own research and that of others they cite, they develop models of different expected patterns based on contrasting ecological processes (e.g., competition between trees will show one pattern, whereas microsite variability will result in another pattern), and then they test which model most closely fits the observed patterns. Likewise, Erica Fleishman and colleagues show us in Box 7.2 how they used observations of bird distributions to test and refine models of theoretical species distribution patterns. Ultimately, these authors remind us that we shouldn't confuse biologically impenetrable patterns—those things that are just naturally too complex for us to understand completely—with analytically impenetrable patterns, which are just problems waiting for different, or better, ways of examining them.

A second source of skepticism regarding observation-based studies is that many take what seems to be an inductive approach to achieve understanding. That is, they build up layers of data to put together a plausible narrative to explain an ecological phenomenon. Inductive inference is often argued to be weak, as illustrated by the "black swan" metaphor. The idea is that if you lived in Europe before Western naturalists had been to Australia and you observed swans for years and years you would still only see white ones. You might then inductively come to a belief about swans such as this: "Every swan I have observed is white, therefore all swans are white." But when explorers set foot in Australia and found black swans, they could instantly falsify the notion of all swans being white—a clear victory in terms of efficiency and accuracy of the deductive approach over the inductive (Taleb 2007).

In reality, this is a philosophers' game. A real naturalist, a real ecologist, or any real scientist doesn't think like this at all. If you discovered a flock of a new species of bird on a remote island and they were all blue, you would duly note it and suggest that blue was their identifying color, along with numerous other physical and behavioral characteristics that you carefully recorded. If a red one suddenly showed up a week after you published a detailed monograph on the species, it might be a bit embarrassing, but it wouldn't be a devastating refutation of your core beliefs or even a reflection of your inadequacies as a scientist, but simply a new observation providing more data about a curious phenomenon. All good scientists do well to remember that their theories and ideas are subject to refutation at any time. Only if the goal of science was to search for absolute truth would the "black swan problem" be a real problem. In fact, the extreme complexity of ecological systems will always surprise us whether we use observational, experimental, or modeling approaches.

Another problem with singling out induction for criticism is that it is actually everywhere in scientific reasoning, and in fact, most reason-

**BOX 7.2**

## Models of Species Distributions Based on Observational Data

**ERICA FLEISHMAN, BRETT G. DICKSON, STEVEN S. SESNIE, AND DAVID S. DOBKIN**

The persistence of most animal populations varies in part as a function of the amount and configuration of their habitats. We define *habitat* as the abiotic and biotic resources and conditions that facilitate occupancy and persistence of a given organism (Hall, Krausman, and Morrison 1997) and as spatially and temporally explicit, with multiple attributes that can be modeled, mapped, and related to species occupancy across large extents (Fretwell 1972).

Statistical models are the fundamental inferential tools used to estimate quantitative relationships between animals and their habitat. Among the models that have become popular for inference across large areas are maximum entropy (Phillips, Anderson, and Schapire 2006), genetic algorithm for rule-set prediction (Stockwell and Peters 1999), and ecological niche factor analysis (Hirzel et al. 2002). Statistical models are most likely to be accurate if they incorporate robust measures of occupancy determined from field surveys. Occupancy can be defined as the expected probability that a given site is occupied by the species (MacKenzie et al. 2006). Whether a species is recorded as present or absent in a given location is affected by the probability of detecting the species at a given site if it is present. Assuming a species always will be detected when present can result in unreliable inference (MacKenzie et al. 2005).

We illustrate these points with our models of habitat and occupancy of two species of breeding birds in the central Great Basin (Lander, Nye, and Eureka Counties, Nevada), MacGillivray's Warbler (*Oporornis tolmiei*) and Spotted Towhee (*Pipilo maculatus*) (Dickson et al. In Press; Dickson et al. 2009). Data on presence

ing that is assumed to be wholly deductive is actually only possible amid a matrix of observational and inductive approaches. Sir Arthur Conan Doyle's Sherlock Holmes character, for example, is commonly invoked as a model of the power of deductive thinking to solve complex problems. But in order to get to the point of making a few shrewd deductive tests to finger the culprit, Holmes actually relies on his own long history of observa-

and absence of birds came from five years of field surveys. We estimated occupancy, weighted by detection probability, as a function of vegetation structure and composition, which were measured in the immediate vicinity of survey points, and topography and land cover, which were derived from remotely sensed data.

Accounting for detection probability, our estimates of occupancy for MacGillivray's Warbler and Spotted Towhee, respectively, were 18 percent and 30 percent higher than the naïve estimates. The better-supported models of occupancy (on the basis of model selection criteria) of MacGillivray's Warbler always included the proportion of deciduous shrubs (derived from high-resolution digital aerial photographs), whereas the better-supported models of occupancy of Spotted Towhee always included the frequency of shrubs (ground data). Indeed, for MacGillivray's Warbler, the habitat-based model of occupancy with the least support was the same as the model of Spotted Towhee occupancy with the most support. Without observational data, we would not have known that two superficially similar species (insectivorous, shrub-nesting, passerine birds) occurring in the same area (e.g., at survey points or canyons) perceive habitat differently.

Site-level variables measured in the field are rarely included in spatially explicit models of habitat quality, but can increase the accuracy of those models. However, elements of habitat to which animals respond strongly, such as vegetation structure and composition, may not be well discriminated at the spatial, temporal, or spectral resolution of commonly used remote-sensing systems. Thus, models of habitat quality may not adequately represent animal-habitat relationships when only remotely sensed or when digital data layers are used to build them, and especially when remote data are not validated in the field. Model projections that are inaccurate or highly uncertain cannot reliably inform management decisions.

tion of crime scenes and criminals, which has given him skills not unlike the well-practiced ecological observers we discussed earlier in the book.

In the real world, a scientist very quickly, and largely subconsciously, eliminates a large number of possible factors, based on her own or others' observations. What scientists and detectives actually do is not to induce or deduce knowledge dryly and logically, but slyly to "kidnap" ideas from the past—either their own subconscious memories, or the collective memories of past research—and use them to advance new ideas. Thus the term "abduction" has been used to describe the process of (apparently) spontaneously generating hypotheses based on initial observations. The process of abduction is about creating new ideas and explaining observations by making analogies to past knowledge, like Darwin writing about selective animal breeding to help explain his theory of natural selection (Lawson 2010). Again, the complexity of ecological systems (and the humans who study them) makes it inadvisable to wholly subscribe to a single philosophy in order to attain ecological understanding.

A third common critique of observational approaches is that they are only "just-so stories" (referring to Rudyard Kipling's children's tales, which recount the most absurdly nonscientific origins of various animals). In scientific circles, the terms *narrative, anecdote,* and *story* are generally used dismissively to describe data or studies that somehow fail to clear a usually unstated hurdle barring them from the realm of scientific legitimacy. In lectures and in scientific conferences you will often hear a researcher presenting some fascinating new observations defensively provide the caveat, "Of course, these data are largely anecdotal." There are good reasons for some of these concerns. Storytellers can be inaccurate or deliberately deceitful, scientists carry their own biases that may cause them to select some stories over others, and sometimes stories are so entertaining that we want them to be true and are slow to reject them in the face of contrary evidence.

Yet in both the real sense of stories passed down through generations that provide ecological evidence (see Chapter 5) and the figurative sense of data that appear story-like, stories are an essential part of ecology (Cleland 2002; Francis and Hare 1994). Janet Gardner and colleagues note that ecological stories reflect the creative aspect of hypothesis formulation, but what makes these stories or narratives scientific is that they are subject to change pending further observations (Gardner et al. 2007). As Paul Grobstein notes, science "is a continual and recursive process of story testing" (Grobstein 2005). Even the most carefully controlled experiments are built around a story and have a story to tell.

In this view, science isn't about removing all personal traces from the scientific method in an attempt to eliminate subjectivity and be completely unbiased. But instead of an impossible quest for a "view from nowhere," Grobstein advocates a "view from everywhere" that uses the greatest amount of data from the widest range of perspectives to create a robust scientific story (Grobstein 2005). This "view from everywhere" is more achievable than at any previous point in history because of new observational technology, along with the wider acceptance of social science methodologies in ecology, but at least some of its roots arise from an early-twentieth-century collaboration between an ecologist and a storyteller. The ecologist, Ed Ricketts, and the storyteller, John Steinbeck, together sought to describe what they called the "toto-picture" of social and ecological relationships, even as scientists around them were becoming ever more reductionist in their investigations. Ricketts and Steinbeck explicitly recognized their own role as humans in shaping their investigations. At the outset of their famous 1940 ecological expedition to study the Sea of Cortez, or Gulf of California, they noted, "Let us go into the Sea of Cortez, realizing that we become forever a part of it . . . that the rocks we turn over in a tide pool, make us truly and permanently a factor in the ecology of the region" (Steinbeck and Ricketts 1941). This kind of acknowledgment, Grobstein argues, is essential in reducing the subjectivity and turning storytelling into a scientific practice.

Many ecological stories built from observations rely on correlating variables with one another, and this is the focus of a fourth common critique. An example of a correlative study would be interviews with older fishermen that reveal certain years where they remember exceptionally poor fishing, and these might be compared to warm years recorded in the temperature logs taken by ships in the region, and the correlation between the two records might be argued to reveal patterns of past El Niño events. Scientists almost reflexively announce that "correlation does not indicate causation" when responding to this kind of study, but is this still true in the current era, now that we have far more observational data than ever before?

There are good reasons to be skeptical of correlations. Misinterpreted correlations can have far-reaching consequences. A perceived correlation between vaccines and autism in children—later shown to be unfounded—has spawned a large movement against vaccinating children, even though the benefits of vaccination to individuals and society far outweigh the potential risks. Moreover, almost any two variables can be correlated. A tongue-in-cheek graph showing a tight correlation between "Number of Pirates vs. Average Global Temperature" was widely distributed on the Internet. It even ended up on the websites of climate-change deniers, who used it to argue that any correlation between temperature and another factor (such as greenhouse-gas concentration in the atmosphere) is likely to be spurious.

But even in the halls of serious science, if you go to enough seminars on ecology you will inevitably see the presentation of correlational data that the author claims have a "significant" relationship in a regression analysis, but that look like a bunch of scattered points with a line drawn through them. Part of the problem is that the statistical tests typically used for correlation tests are not very conservative, meaning that you may get a statistically significant result, even if the underlying relationship doesn't mean much in nature. An amusing exchange of figures in the esteemed journal *Science*, highlighted by several authors on the scientific method as a warning against unquestioningly accepting regression results, occurred when Entomologist David Roubik (Roubik 1978) published the following graph showing the relationship between the number of stingless bees and Africanized bees in a 1978 *Science* paper (Fig. 7.1).

To which Robert Hazen published a succint response: "The rather fanciful curve fitting of Roubik has prompted me to propose an alternative interpretation of his data," which was accompanied by the following modified figure (Fig. 7.2) (Hazen 1978).

In his defense (he seemed to take the "stinging" critique in good stride), Roubik admitted that perhaps a computer should be used in curve fitting, but we are now learning that this can actually make the problem worse, as the most basic computer programs can fit a curve to your data with a single click, but these curves may be no more meaningful than the flight of the bee through Roubik's data.

**Figure 7.1**

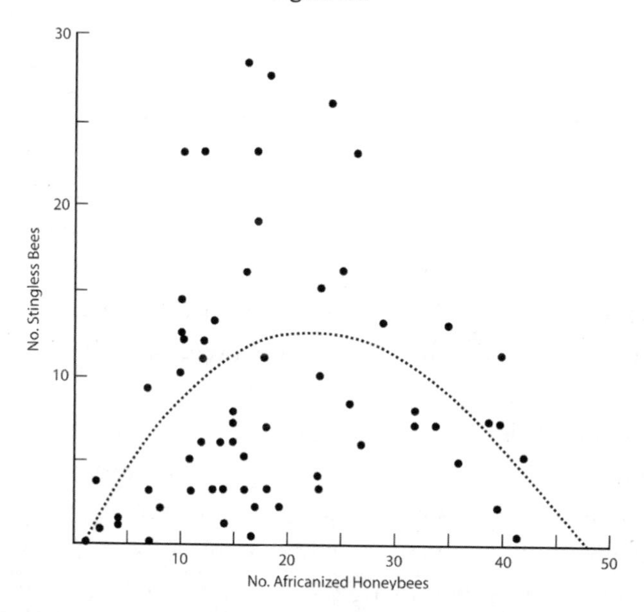


*Adapted from Roubik's plotted relationship between stingless bees and Africanized bees.*

**Figure 7.2**

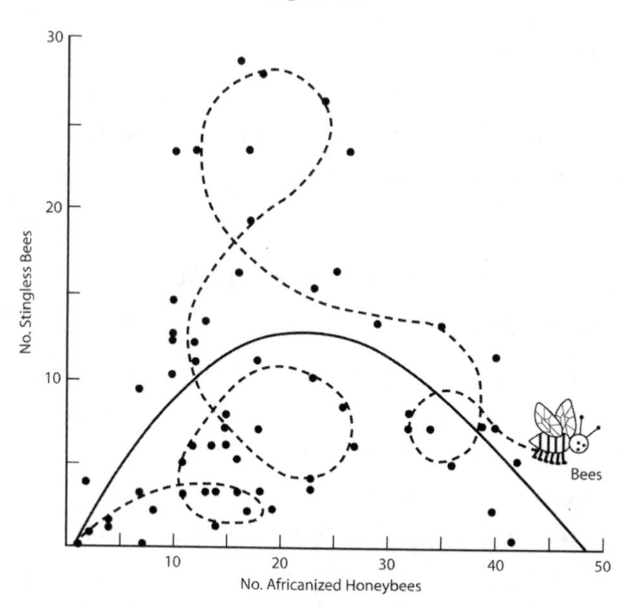


*Adapted from Hazen's alternate interpretation of Roubik's data.*

Nonetheless, it is simply not true that correlated data cannot lead you to the mechanisms underlying a given phenomenon. The trick is that with multiple layers of data it is sometimes possible to ascribe causation to a set of correlations. The story of the massive K-T extinction event that brought the end of the dinosaurs is an especially elegant example of correlating many variables to determine the mechanism—a large asteroid impact and associated global cooling—that brought the demise. Here a highly interdisciplinary team that included Nobel Prize–winning physicist Luis Alvarez and his geologist son Walter correlated stratigraphic data, fossil data, and chemical data—most notably a thin layer of the rare element iridium found at the same level in multiple strata around the world—to test multiple hypotheses related to the extinction event (Alvarez et al. 1980). In keeping with our contention that observational studies often lead to serendipitous results, the Alvarezes were not even looking for evidence of the K-T extinction when they started, but rather trying to develop a geological clock (based on the deposition rate of rare elements like iridium) to time sedimentation rates (Alvarez and Asaro 1990). Their work is all the more remarkable because they did not at the time have the "smoking gun" of a known impact crater large enough to be associated with the events they inferred. That crater would be found 10 years later, deep below the surface of the Yucatan peninsula, using previously unavailable observational technology. Correlation may not always imply causation, but in this case the correlations seem to have been strong enough to wipe out the dinosaurs!

### Integrating Observations and Other Means of Achieving Scientific Understanding

The dismissal of observational approaches in ecology as a valid scientific tool has percolated into research-funding agencies, journals, and academia in general. This may have limited the search for novel and creative ways to help us understand ecological systems, and it diminishes the importance of past and current knowledge based on observations. A "system sufficiently simplified" is increasingly difficult to justify when we need to understand the dynamics of complex natural systems that are intricately tied to complex human behaviors.

But nor can observational studies get us to this understanding alone. As James Quinn and Arthur Dunham conclude in their much-cited critique of formal hypothesis testing, "Healthy skepticism toward a single methodological model seems thoroughly as appropriate as toward any other claim of scientific truth" (Quinn and Dunham 1983). Likewise, Robert Vadas Jr., using the story of a scientific controversy over the meaning of so-called honey-bee foraging dances as a lesson, argues that a mixed approach in which both verification and falsification are used is necessary to reach conclusions in complex ecological systems (Vadas 1994). And although we've used Popper's most noted philosophy as a straw man, we have to acknowledge that he himself actually advocated a pluralistic view that wasn't bound by past dogma or traditional approaches (Walker 2010). In other words, just as scientists should question any scientific conclusion, we should also question any methodology. Maintaining this ever-questioning attitude is likely to go much further to ensure that a given ecological study brings us closer to understanding than any prescribed method of how to do science.

Fortunately, both long-deceased and contemporary scientist-philosophers have elaborated viewpoints and methodologies that are inclusive of the wide range of approaches needed to deal with today's complex social ecological systems. The brief survey of these works that follows gives the sense that there are many ways, philosophical and practical, to reconcile reductionist and holistic viewpoints as well as experimental and observational methods.

Steward Pickett and colleagues' "Ecological Understanding" (Pickett, Jones, and Kolasa 2007) presents a readable ecological philosophy for the twenty-first century that explicitly acknowledges the balance needed between hypothesis-driven experimental research and speculative observations. Methods for putting these inclusive philosophies into practice are also emerging throughout the sciences. Glenn Suter and others advocate a "weight of evidence" approach toward determining causation, which has appeared in various forms in epidemiology and in environmental impact studies (Beyers 1998), both fields where controlled experiments and strictly falsifiable hypotheses are difficult. In this approach, multiple lines of observed evidence are integrated and weighed against one another based on previous studies and knowledge of the system. The "weight of evidence" approach is amenable to studies that combine observational and experimental approaches as well as observational studies on their own. As Suter describes it, "confidence about causation is obtained by finding concordance of observational results with the results of controlled studies and consistency among observational results at different sites" (Suter 1996, p. 344). Although Suter concedes that this approach will appear less objective than strictly deductive hypothesis testing, he suggests it is more likely to be correct because it uses a much wider range of data.

C. S. Holling and Craig Allen advocate a cyclical process of pattern finding and tests to determine causes of the patterns, which they call "adaptive inference." "Adaptive inference," they write, "relies on the exuberant invention of multiple, competing hypotheses followed by the assessment of carefully constructed comparative data to explore the logical consequences of each" (Holling and Allen 2002). Along the same lines, Judi Hewitt and colleagues proposed an approach integrating observational and experimental manipulations, beginning with observational natural history to identify the likely scale of the problem, potential causal variables, and feedbacks. Depending on the scale and complexity of the problem, experimental manipulations are either built into a large-scale observational/correlational framework or applied in an alternating fashion with correlational studies, with each type of study providing information to better focus the next iteration of the other (Hewitt et al. 2007).

There are good examples emerging of studies that integrate observational and experimental approaches. Using the example of a reed invasion into farmed salt hay fields, David Bart (2006) showed that local ecological knowledge—which often provides excellent historical observational insights but little power to resolve causal mechanisms—could be strengthened with experimental manipulations. Joanna Norkko and colleagues (2006) showed that integrating physiological indicators honed in controlled laboratory experiments with broad population sampling across ecological gradients can elucidate linkages between mechanisms and ecological patterns and address questions such as "What is the cause of a species range limit?" and "How will this species respond to climate warming?" Some experimental manipulations have now been carried out for a long-enough period of time to be combined usefully with observational data in order to resolve complex hypotheses such as the relative role of natural versus artificial selection (size-selective harvesting due to fishing) in driving evolutionary change in fish (Coltman 2008; Coltman et al. 2003).

Although there has been a recent concentration of arguments for more-inclusive forms of scientific inference, the ideas underlying them are not modern inventions. Science philosopher and evolutionary biologist Massimo Pigliucci notes that the nineteenth-century philosopher William Whewell used the term "Consilience of Induction" to describe a process by which layers of different evidence taken from a wide range of viewpoints point toward a similar answer, as in the Alvarezes' K-T extinction study (Pigliucci 2002). Rather, we are seeing a greater attention to more-inclusive ecological philosophies because they reflect the reality of how we need to achieve ecological understanding in the twenty-first century.

So, are observational approaches scientific? That question can only be answered by the studies themselves. Our contention throughout this book is that the strength of observational approaches will increase by incorporating a wide variety of data types, data gatherers, and strategies, but that their ability to be useful to science depends on the context in which they are used. Finding the science in observation is about finding the meeting place of the spirit of discovery that is shared when one keen observer of the world passes on an ecological story to another, and the ability to achieve a rigorous ecological understanding of a particular ecological phenomenon. Strong conceptual frameworks, large amounts of data, and the right analytical tools are essential if observational approaches are to contribute to ecological sciences. Demonstrations of the rigor within observational studies are amassing within ecological science, especially as more data and better analytical techniques become available.

In the next part of the book we show why it is so important to get to the science within the observations. Having the backstop of rigorous science is increasingly essential as observational studies themselves are being increasingly called upon to contribute to resource-management and policy debates, as we discuss in the next chapter, and it becomes critical as observational methods are infused with new educational models at every level, from kindergarten to graduate school, as we discuss in Chapter 9.

PART IV

# BEYOND ACADEMIA: THE POWER OF OBSERVATIONAL APPROACHES

Part IV is the payoff for us. Here we show how all the components of observation-based approaches to ecology—the different types of observations and observers, the new and old analytical approaches, the philosophical consilience of different ways of gaining ecological understanding—come together to play an integral role in the things we really care a lot about: environmental policy and environmental education.

In Chapter 8 we consider the policy-making process from an ecological point of view, trying to elucidate the holistic picture of what influences policy change. In this view, it is not enough to convey the technical information behind a particular policy prescription, but further attention must be paid to emotional and sociological aspects of policy making. We argue that observational studies, more than experimental or theoretical approaches, are easy to translate into technical, emotional, and sociological content that can be used by policy makers. In Chapter 9, we argue that observational approaches are amenable to educational opportunities at every level and in both formal and informal settings. We show that both the process of gathering observational data and the products of those data can be made into lifelong lesson plans that create positive feedback loops that, in turn, encourage further nature education, scientific exploration, and advocacy for environmental protection.

CHAPTER 8

# Ecology's Renewed Importance in Policy

We have been hinting all along that ecology has a more prominent role to play in public policy debates than ever before. The multiple-scale and large-scale ecological changes that are tailored to large, multiple-scale networked ecological observations that we discussed in Chapter 6 should also be the focus of policy changes at multiple scales of governance—from local to global. Yet career ecologists have been frustrated at how little progress ecology as a science has made in turning the tide of environmental degradation and destruction. In this chapter we dive into this paradox by looking at the ecology of policy making itself and identifing matches and mismatches between ecological science and the complex ecology of politics.

To put the problem bluntly, policy makers don't read *Ecological Monographs.* Although more and more ecologists are conducting work that should be relevant to policy makers, there is a wide gap between the practice of ecological science and the practice of politics. Ecological studies, whether or not they are intentionally initiated in response to a policy or management need, will increasingly have a role to play in the policy-making process and they can only do so if they can be brought across the science-policy divide. In this chapter we identify the sources of this divide. We highlight examples of past successes of observational ecological studies in influencing critical environmental policy issues, and we discuss why they fill a role in making the science-to-policy connection that can't be filled by experimental and theoretical studies. We argue that observation-based approaches to ecology are highly amenable to informing a policy-making process essentially because they are useful in translating the language of science into the language of policy. While this is a promising sign for the future relevance of ecology in helping to solve urgent environmental challenges, it also brings with it the potential to misinterpret or misuse results and to overreact to scientific findings.

Policy has three distinct but intertwined components that must be activated in order to move a policy change forward. There is a technical component that comprises the facts and figures, the calculated risks and benefits, and the demonstrable outcomes of any policy. There is an emotional component that is stimulated by the feelings, ethical considerations, and spiritual beliefs that drive both calls for policy change and also people's responses to policy change. And there is a sociological component that comprises the politics and relationships, personal histories and animosities, coalitions and betrayals that all come into play when a new policy is proposed and debated.

## Policy Making Takes More than Good Data

Changing policy takes much more than just getting scientific information to policy makers, but delivering the information is a critical first step. This connection can be made in many ways: through citizen groups that use science to advance their agendas; through the portrayal of scientific results in the media; and through direct connections between scientists and policy makers. This direct connection is vital, but must be absolutely clear and defensible in order to maintain the credibility of scientists, which is why Stuart Pimm argues in Box 8.1 that it is so important to be a direct observer of change in the world.

In all of these cases, including the direct transfer of information from scientist to lawmaker, there is some kind of translation of the original science into a language that is considered more relevant, or persuasive, to a lawmaker. This interpretation is not just technical, but emotional as well. People and organizations trying to convince lawmakers to change policy

BOX 8.1

## Observation and Policy: The Importance of Being There

STUART PIMM

"I was there," Elrond tells Gandalf in *The Lord of the Rings*. It's a chilling moment, for Elrond has witnessed a previous failure to destroy the evil ring. I have been the witness to failure, too. Experiencing it made me a conservation biologist—before we had the name for our new profession—though not 3,000 years ago, as in Elrond's case.

That experience shapes how I convey messages—to the media, to Congressional committees when I have testified before them, to public groups, and to church congregations.

There's a story of a senior figure from a conservation NGO talking publicly about extinctions and how many there are, then being asked to name them and being unable to do so. Shame! Extinction is a very real experience for me.

Birdwatchers list the species they see, and indeed, some become obsessive about how many and where. The visibility of birds and the numbers of their species mean that an understanding of their diversity and biogeography comes easily and quickly. I had a life list at age 13. When, 15 year later, I headed to Hawai'i, I already had the list of species I needed to see. I was doing fieldwork, six days out of every eight, and for months. I was convinced I would see them all—even the rare ones.

I didn't.

Some of the species I sought were already extinct; some were exceptionally rare. I was a skilled observer and no matter how much time I spent in the field, they weren't there.

I learned other lessons too. Species with small ranges are the ones most likely to be locally scarce—a double jeopardy that makes them disproportionately at risk of extinction. The places where such species occur are idiosyncratic and, surprisingly, they are not where the greatest numbers of other species live.

These are the "laws" of biodiversity and they shape conservation priorities. So, yes, when asked I do know where the recently extinct species are, sometimes from having seen them. And I do know where the species that teeter on the brink are to be found. And, well yes, tropical deforestation is personal too, for I have stepped far too often into the black ash of what was once a forest teeming with life. ▸

> ▸ I was there, I saw—and see—the mistakes that lead to a loss of biodiversity, and this shapes both my science and, importantly, how I communicate it to the public and to policy makers.
>
> I witness the successes, too: grey and blue whales out in the Pacific, the peregrines that migrate in the hundreds over my home in the Florida Keys each fall. When challenged about the efficacy of the Endangered Species Act in the USA, yes, I can testify that when we put our minds to it, we need not witness failure.

need at some level to get the facts straight (though anyone watching policy debates can see how far those facts get stretched at times), but most policy change is driven by some visceral attachment or response to an issue. This is why adorable polar bear cubs and not, say, intertidal snails have become the poster children for environmental groups' campaigns about climate warming.

There has been a lot of attention paid to "science communication" in recent years in the form of science blogs, articles, books with titles like *Don't Be* Such *a Scientist* (Olson 2009), and even "training programs" (e.g., the Aldo Leopold Leadership Program—see leopoldleadership.stanford.edu). These have been enormously important, but they have largely focused on the technical and only a little on the emotional gaps between science and policy (Groffman et al. 2010). The best of these efforts try to move beyond the rather offensive advice that scientists need to translate their work into a language that the "general population" can understand, or worse, that scientists should simplify their language to a "fifth-grade level," but they still tend to coalesce around the theme that scientists are nerdy (or arrogant) technocrats who could be infinitely more effective shaping policy if only they could learn to communicate. This view ignores the real ecology of the political process. The fact is that there are many conscientious scientists out there who communicate about their work very clearly, accurately, and without condescension. As Rafe observed during his time working in the U.S. Congress, lawmakers and their staffs are actually inundated daily by a barrage of newspaper clippings, articles, reports, white papers, podcasts, e-mail blasts, and lobbyists' spiels that largely get the science right and communicate it clearly, but just aren't that useful to the policy making process.

What is needed, even more than simply better communication, are ways to translate urgent science into language that resonates in all three areas of policy making—the technical, emotional, and sociological. Experimental and theoretical studies of ecology, whether designed in response to a policy need or (more commonly) not, simply aren't able to do this because they operate at small scales of space and time, or in a computerized world that can't be touched or smelled or seen in reality. This is not to say that they can't produce results relevant to policy, but only that they have a harder time making their case for relevance to those involved in the policy-making process because they don't resonate strongly in the three areas of policy.

Observational approaches to ecology, by contrast, tend to produce results that translate very well in the technical and emotional components of policy making, and the recent expansion of ecology toward creating observational data sets in conjunction with citizen scientists, as well as the local and traditional knowledge holders whom we discussed in Chapter 5, creates the kind of coalitions of interest that can influence the complex sociology of policy making. Likewise, the opening of ecology toward new fields in social sciences, such as consumer behavior, is more directly connecting ecological results with an understanding of how and why people demand political change.

Experimental studies can be useful for understanding *mechanisms* behind ecological change at small scales. Policy makers, by contrast, are typically interested in *outcomes* of ecological change at large scales. It can be hard for them to see the connection from what happens in a small number of plots in a nature reserve to how that will affect people across the much larger areas where they have the power to make public policy. Even attention-grabbing headlines from experimental work, such as a recent finding from the Duke University experimental forest showing that increased carbon dioxide (when added in an experimental setting) led to faster-growing and more-toxic poison ivy plants (Mohan et al. 2006), leave many unanswered questions that make it difficult for policy makers to latch on. For example, would poison ivy really grow like this outside the Duke Forest in North Carolina? Would poison ivy really adapt like this given that the rate of increase of carbon dioxide in the real world is much different from the high concentrations of $CO_2$ suddenly present in experimental forest plots? Would some other weed not even considered in the experimental setup thrive and outcompete poison ivy in the real world? These questions can't be addressed at the time and spatial scales of experiments, which is why scientists at Duke's experimental forest have always insisted that their results are most useful for feeding into ecological models that can get at those complex issues (Dellwo 2010).

Modeling—creating computer-generated simulations of future ecologies—can help us understand the changes likely to come in the future, and models can be applied to large spatial scales. As computing power and ecological knowledge has expanded (in part through experiments like those at Duke Forest), models have become increasingly sophisticated, bringing in more interacting variables and, we hope, coming closer to the ideal situation where they can both accurately "hindcast" the past (that is, re-create past observed conditions to show that the models get the ecosystem dynamics correct) as well as forecast the future. Ecologists use all sorts of models to look at the likely spread of invasive species, the risk of fire outbreaks in different locations, and the projected future populations of threatened, endangered, or commercially exploited species. When they work, they can provide a plausible range of future conditions that may occur given different assumed changes in policy or management.

But no matter how sophisticated they get, models will always suffer from a widespread human prejudice against computers. Although many of us use and rely on computers for our very lives (when we fly in commercial airplanes, for example), in general, people are skeptical of computer-generated models. This is probably unfair and it's really no fault of the modelers, who have continually improved climate, population, hydrological, and even human decision-making models. It's just quite simply that not seeing is not believing.

Some of the skepticism about models will be broken down as more observations confirm or correct them. For example, over 100 years ago

Svante Arrhenius relied on very simple models to make the first predictions about human-driven climate warming. While his theories were known to a small group of scientists throughout the twentieth century, they didn't spread very far because they lacked the observational validation that would attract more scientists to consider the accuracy of his predictions or that would convince the public that there was something real to be concerned about. But through the dogged determination of Charles Keeling, who developed and deployed an ultra-sensitive monitor for atmospheric carbon dioxide, we would finally have a more accurate picture of greenhouse-gas accumulation (Gillis 2010). The data would confirm that Arrhenius was theoretically correct, though factually far off the mark about how quickly carbon dioxide would accumulate in the atmosphere. More important, the so-called Keeling Curve would become a startling and incontrovertible observational picture of the legacy of industrialization and its potential consequence for global ecology.

The curve illustrates both the overall rise of $CO_2$ concentrations in the atmosphere and also a seasonal signal of planetary carbon balance in the form of small jagged up-and-down oscillations in the curve. If you focused in on just a few measurements from the curve—as Keeling must have done when he first started getting data from his sensors—you would just see the concentration of carbon going up and down throughout the year. It wouldn't be until several years of observation that you could identify this oscillation clearly as a seasonal cycle with $CO_2$ levels rising during northern hemisphere winter (when trees spread across a larger landmass than that of the southern hemisphere have slowed or ceased their uptake of carbon), and falling during summer. And it would take a decade of observation to establish that an increase in $CO_2$ was occurring above and beyond the seasonal spike.

It is this clear picture that emerges, simply wrought from years of observations and accounting for several scales of time (seasonal, yearly, decadal), that makes the profoundly important connection between the prescient theory of Arrhenius and the more accurate and sophisticated supercomputer climate models that are now used to project the future state of our world. The curve (now yielded repeatedly from $CO_2$ monitoring stations throughout the world) is the solid backbone in the huge body of climate-change research that can seem amorphous and unwieldy to anyone who doesn't study climate change for a living.

Modelers have always acknowledged that observations are essential—both historical observations that help them build the models, and present-day observations that help validate previous outputs from the models. But observations also have a role in reassuring the public that the outputs of a silicon-based machine can tell us something about the real world. Observations do this by providing both technical support, in the form of data that link the quantitative and qualitative predictions of models to actual changes in the real world, and also the emotional support which shows that our world isn't just changing in the pixels of a computer screen, but in the lives and fates of things we really care about. Models can do many things, but they rarely convey the emotional content of environmental issues. Rafe jokes that his grandmother doesn't call him up concerned about the latest computerized climate-model outputs from the World Climate Research Programme, but she does call about the polar bears haplessly swimming between ever-scarcer ice floes in the Arctic that she saw on the Discovery Channel.

## The Emotional Power of Observational Approaches

This technical and emotional power of observational approaches to influence public debate has been seen throughout the history of ecology. When he was president of the United States, the naturalist Teddy Roosevelt used photographs of deforested and eroded Chinese mountains during a Congressional address (Cutright 1985) to make the case for a national scientifically managed forest service, an idea that became one of the most important conservation legacies in American history. There were both technical and emotional aspects of Roosevelt's entreaty, but given his penchant for dramatic oratory, we suspect that, for Americans in the throes of industrialization, the technical story of erosion following deforestation was secondary to the emotional appeal of showing how science could improve society.

Just as scientific results can come from observations filtered through all our senses (Chapter 3), the emotional content of these observations is not only transferred through our visual field. Rachel Carson's enormously influential book *Silent Spring,* which is credited with launching the modern environmental movement, used the auditory sensation of silence in places where birdsong used to fill the air in order to galvanize public concern about pesticides such as DDT. In the southeastern United States it is the *smell* of hog farms, rather than the more distal concern about methane-gas emissions, that has created alliances of local citizens, scientists, environmentalists, and hog farmers themselves to develop ways to capture the methane and turn it into fuel.

Because they play on our emotions, even small slices of larger observations can have disproportionate effects on policy. In 1987 the surreal image of a garbage barge circling Manhattan with no place to dump its cargo became a powerful symbol and catalyst for massively renewed efforts at recycling (Miller 2007). Public response to fires on the Cuyahoga River in Ohio in 1969 are considered to be the genesis of the U.S. Environmental Protection Agency and the passage of the Clean Water Act in the United States (Adler 2003), which both have influence far beyond water quality in a midwestern river.

Rivers catching fire and excess garbage circling a city are fairly localized environmental issues that sparked an environmental consciousness that reverberated across the world to places that experienced similar localized problems. A more profound awareness that observations have provided is the understanding that our actions can actually alter global dynamics unilaterally, not as the additive property of the same mistake repeated in different places, but as a comprehensive attack on the whole system. This awareness didn't emerge automatically, but required a priming step in the form of the compelling story of the K-T extinction event described in Chapter 7. The dinosaur extinction story's immediate impact was to increase fear about asteroids rather than raise awareness about our own impacts on global ecosystems, but it made the case conceptually that our entire Earth system could shift from being a life-supporting to a life-destroying planet.

The awareness that we as a species could be a driver of this shift would come just a few years later when the popular scientist Carl Sagan and colleagues postulated that an all-out nuclear battle between the United States and the Soviet Union, which was a very real and frightening prospect at the time, would lead to such expulsion of dust and soot from fires into the atmosphere that the entire planet would suffer a "nuclear winter" in which solar radiation would be so reduced that the vast majority of macroscopic life-forms on Earth would die (Turco et al. 1983). What made this conjecture and the authors' associated models seem plausible in people's minds was the previous observations of the K-T extinction. The mechanism of extinction was remarkably similar, only the delivery vehicle—an intercontinental ballistic missile vs. an asteroid—was different. For the first time, large numbers of people could comprehend the once-incomprehensible—that humans were capable of causing global-scale disruption of our living systems.

It was then only a few more years until our awareness of our global impact was stoked yet again. In 1985 Joseph Farman and colleagues published observational evidence that a large area of the stratospheric ozone layer, which protects life on Earth's surface from excessive UV radiation, had declined dramatically in concentration (Farman, Gardiner, and Shanklin 1985). The result that Farman and his colleagues displayed should have been a surprise to no one. After all, the chemical model demonstrating that chlorine and associated atoms in the upper atmosphere could catalytically destroy ozone was published way back in 1974 by Mario Molina and F. Sherwood Rowland (Molina and Rowland 1974). But the model and the raft of scientific studies following the 1974 paper failed to captivate the public in the same way that a simple image of the result was able to do. Only two years after the images of Farman's "ozone hole" were published, the Montreal Protocol, a binding international agreement to phase out ozone-depleting chemicals, had been passed, and it remains the most successful effort at global environmental problem solving to date. Rowland and Molina's continual advocacy for a ban on ozone-depleting chemicals was certainly vital to this effort (Meyer et al. 2010), but even according to Molina, the explosion of public attention following the ozone-hole images was essential to achieving the final policy (M. Molina, pers. comm. to RDS, June, 2009).

## Getting Scientists and Citizens Involved

In most cases, communicating science in a way that influences policy takes more than clever framing of technical concepts (Sagarin 2010), and it takes more than an appeal to emotional responses, which can be powerful, but can just as easily (perhaps more easily) be brought to bear on policy that pointedly opposes scientific findings (such as teaching "intelligent design" or arguing that protecting endangered species will destroy jobs). Making lasting change also requires getting into the complex networks of relationships of power and history and shifting motivations that make up the sociological component of policy making. In other words, we need to understand the ecology of politics in order to bring ecology into policy.

Getting scientists directly involved in the process of policy making is an effective, but limited, way of making this connection. There are some opportunities for ecologists to serve as science advisors in national and international government offices, but these relationships are usually short-term arrangements designed mostly to serve as learning experiences for the scientists. At the level of career politicians, other than U.S. President Teddy Roosevelt, rarely has someone with both direct knowledge of the technical aspects of ecology and a deep emotional connection held such a powerful political office. Few ecologists will grow up to become leaders of a country and many of us wouldn't want to. Moreover, as political campaigns become ever more expensive (a recent gubernatorial race in the state of California cost $250 million, with the losing candidate spending $140 million of her own fortune), scientists, who have rarely amassed fortunes, are increasingly unlikely to hold high public office.

But as the upper end of politics is closing its doors to public participation, the bottom end is becoming more open, with citizens more directly involved in meeting local social and environmental challenges. Businessman and activist Paul Hawken has documented the remarkable recent success of small, localized action groups relative to the failures of governments and large international conservation organizations in achieving social justice and environmental protection (Hawken 2007). The opening of ecological science that we have discussed so far creates ripe opportunities in two directions—with citizens getting more active in the process of ecological science, and the process of ecological science getting more involved in the lives of citizens—that will both influence policy making.

The most direct way this will occur is through the continued involvement of citizen-scientists in ecological studies, which connects people back to nature and allows them to tap into the same emotional source of inspiration that makes natural scientists excited about what they do (Revkin 2010). Such involvement can't be mandated by policy; rather, it must be something that people come to naturally to fulfill curiosity or to feel like they are contributing something of value to their local ecology or to a place they visit. Creating opportunities and reducing barriers to this involvement is essential. The National Park Service in the United States, created top-down by the naturalist-president Theodore Roosevelt, is now using a bottom-up approach to gaining support for their continued existence. It has created a series of highly publicized "Bio Blitzes" in which teams of citizens and scientists converge on a single national park over the course of a weekend and record all the plants and animals they can find. The goal is to create involved citizen-observers who discover the value of the parks and the value of ongoing nature observation.

But asking for more engagement of citizens in science can be matched by asking more of scientists in engaging with civic life. This means, in part, engaging the social sciences increasingly within ecological science. This will expand the subjects of our fieldwork from starfish and pine trees to human users and consumers of natural resources. Areas ripe for such work arise where complex ecological systems meet complex economic and policy choices. For example, a growing body of research is being conducted on developing "eco-labels" for fishery species, much like an "organic" label for produce. The point of such labeling schemes is typically to reward and encourage sustainable fishing practices by increasing access to markets such as Whole Foods or Walmart that have stated they want to sell sustainably harvested foods, and by providing a price premium to producers. While a lot of typical ecological monitoring and population surveys are necessary in order to consider whether it is feasible to certify a given fishery as sustainable, we have also discovered how important fishermen's attitudes and beliefs, as well as consumers' preferences, are in this complex equation.

While still a master's candidate at Duke University, Wendy Goyert took on the daunting task of interviewing Maine lobstermen at docksides throughout the coast to determine their opinions and concerns about a statewide proposal to list Maine lobster as a Marine Stewardship Council–certified fishery (Goyert, Sagarin, and Annala 2010). She also conducted a consumer-preference survey to see what drives buyers' decisions in selecting seafood. What she found was that attitudes about the certification program were almost perfectly split among lobstermen between those who liked the idea and those that hated it. Moreover, her consumer survey showed that rather than environmental factors, consumers care much more about freshness and the location where their lobster was caught. She concluded that more benefit would accrue to lobstermen if a label on their product identified it as "Fresh Maine Lobster," with details on the particular location and people who caught it, than if the label identified its sustainability, even though the same fishing practices would go on behind the scenes. In other words, by correctly identifying the human behavioral drivers in a complex system, we can find ways to reach the same end (sustainable fisheries) by unexpected means (a consumer label that doesn't focus on sustainability at all).

An ancillary benefit of this kind of research is that it begins to break down the divide between academic scientists working with natural resources and people who make a living off consumption of those resources. Goyert found that although Maine lobstermen have a reputation as provincial and suspicious of outsiders, they were happy to get a chance to talk with her about a political process that they felt they had little control over.

Breaking down this sense of disconnection between resource users, scientists, conservationists, and policy makers will go a long way toward creating better conservation policy. They are all essential players in managing what are collectively known as "public trust" resources—those ecosystem goods and services that do not fall under direct ownership rights. The United States and many other countries are actually guided by a "Public Trust Doctrine" which acknowledges that natural resources such as wildlife, shellfish beds, and coastal access cannot be owned by a government, but can only be held in trust by the government for the good of all citizens, in the present and for future generations (Turnipseed et al. 2009). In most countries, the public trust doctrine is a matter of common law, rather than a written part of individual environmental policies, meaning that for it to take effect citizens must actively assert their right to have public-trust resources protected on their behalf. This can happen when concerned citizens rally to have a particular aspect of the public trust protected by law or when they sue government for a breach of its public-trust responsibility. But such citizen-focused stewardship can only occur if we know what is in our public-trust "portfolio"—the collection of natural assets shared by all people. Proper and active management of a public-trust portfolio requires learning from and balancing the needs of a mutually dependent network of resource users, lawmakers who must carry out the government's duty to protect the trust, and scientists, citizens, and ecological knowledge holders working to get a clear picture of the resources within that trust. We may never be able to make the "full accounting of his stewardship" required by law of a financial trustee (Scott 1999), but increasing our observational knowledge of the natural resources held in trust will certainly help improve the ability of governments to fulfill their trust responsibilities.

We have argued that changes in the environment have been forcing changes in the way we conduct ecological science. Ideally, these changes should in turn feed back to drive changes in policies that affect our environment. This won't happen automatically because most ecological studies are not designed at the outset with the goal of influencing policy. Rather, observational approaches to ecology are particularly powerful in policy discussions, in part because they naturally operate at the same scales as the images that capture public attention. Rowland and Molina's model applied to individual molecules, incomprehensibly small. But Farman's observations were on a huge scale—a looming hole threatening to swallow up larger and larger parts of the whole Earth—and public perception of the problem was likewise enlarged. Observations come in a language that can be understood globally with little need to translate.

Indeed, many of the headline-grabbing science news items—the ones that do get the attention of lawmakers and their staffs—are observational in nature. A distant solar system captured by the Hubble space telescope. A strange new squidlike animal caught in the floodlights of a deep-sea submersible. A massive dead zone in the Gulf of Mexico revealed by satellite images. To be sure, these fascinating snapshots are not necessarily scientific at all, but rather the superficial manifestation of extensive scientific research occurring before and after the images are published. Such images, when broadcast widely, can have profoundly cascading effects on all sorts of social decision making. They ignite public debate and catalyze greater funding for space exploration, or biodiversity protection, or new water-protection legislation.

Because these symbolic images are powerful, they may cause us to overreact or may distract our attention from more important issues. The validity of these images in terms of reflecting a larger reality and also their power in attracting continued activism both depend on how well they truly match the larger scientific enterprise that produced them. Had the Hubble telescope only produced a few pretty pictures, interest in the science behind it would have waned, along with the strong public support for continuing its mission. Likewise, if the first images of dead zones had only shown a freak phenomenon, never to be repeated, rather than an ongoing and growing problem that later turned up in other parts of the world, it wouldn't attract much attention about water-quality issues and may have even been held up as yet another example of exaggeration by environmentalists. It is the linkage between fleeting but dramatic images and the less glamorous long-term science underlying the images that establishes a strong nexus joining science, society, and policy.

Observations may serve as a proxy for larger, more-complex issues. In this sense, starving polar bears are a good proxy for climate change. They show that climate change is affecting something that many people care about (in this case, animal welfare) and that climate change is happening now, not 50 years from now. In other ways, of course, polar bears can only capture a tiny subsample of effects that climate change will bring, and little of the complexity (for example, there may be different populations of polar bears that fare better in a climate-warmed world).

This potential for observations to both fairly and unfairly influence policy debates puts added responsibility on the scientist, whether she intended her study to respond to a policy need or not. Results of observational studies are increasingly transferred into news stories and documentaries and then discussed on blogs and shared across social networks. Once outside the safe confines of a peer-reviewed journal, they can be used in all sorts of ways to influence policy. This raises key questions about where the scientist's responsibility ends. For example, Rafe was surprised to find himself involuntarily added to a list created by the conservative Heartland Institute of "500 Scientists with Documented Doubts of Man Made Global Warming Scares." Although the Institute never said why he was on the list, presumably it was due to a short article he published in *Nature* that identified a small but correctable error in the way all published studies had documented shifts in phenological timing (Sagarin 2001). Of course, pointing out errors in published data is exactly how climate science or any other science progresses toward understanding, and the fact that the correction was published in the leading scientific journal, rather than buried in an online blog or conference proceedings, is testament to the fact that scientists take their self-policing role quite seriously. But does the scientist have a responsibility to correct an individual or organization that sidesteps any review process and grossly misinterprets his science?

In this case, the infamous list was quickly denounced and shown to be fraudulent by a public-interest group focused on the integrity of science (desmogblog.com/outrage-in-the-climate-science-community-continues-over-the-500-scientist-list), but what if it had been taken seriously by mainstream media or influential lawmakers? Deciding when to step away from the relative safety of academia and intervene directly in policy debates is a choice that observational scientists will increasingly need to make. Some scientists have already decided that it is irresponsible *not* to engage in policy debates, considering the damage to ecosystems that has already been done and the relative lack of serious policy response (Meyer et al. 2010; Whitmer et al. 2010). Others consider that such advocacy should only take place in the narrow circumstances where a policy debate concerns something directly related to the scientist's own research. In practice, considering the interconnections between organisms and across scales in ecology, it is very hard to draw that line. We find it almost inevitable that tomorrow's scientists will be drawn into policy discussions. Thus, as we discuss in the next chapter, scientific education going forward will need to be more holistic and inclusive than it has been. We will argue that observation-based studies feed a catalytic cycle of education that broadly trains ecologists who, in turn, become the progenitors of the next generation of observation-based studies.

CHAPTER 9

# Opening Nature's Door to a New Generation of Citizens and Ecologists

Humans by their nature are observers. Before we are even born we are sensing our environment. As we grow older, we acquire crucial information mainly by observing the environment directly or through our peers' interactions with the environment. There is a natural connection between observation and how we learn and understand the world. We have evolved to use observations to build up our own representation of reality. As the American Association for the Advancement of Science and the National Science Foundation have acknowledged, the "practices of observation," particularly "observing nature," are essential to build the ability to apply the process of science in biology (Brewer and Smith 2011).

There is, accordingly, a natural nexus between observation of nature as an educational tool and as a scientific tool, and both branches of this convergence can mutually benefit society and the science of ecology in several ways. First, the use of observation as an educational tool can benefit the psychological and intellectual development of children. Second, early and memorable experiences with nature create children, adolescents, and young adults who grow to appreciate nature and will continue, either formally as scientists or informally as citizen knowledge holders (see Chapter 5), to join the ranks of ecological observers. These observers are then more likely to become those who make the connections between societal actions and ecological change. Third, at higher levels of education, nature observation can give students the inspiration to pursue careers in the life sciences as well as the more immediate benefit of giving them sources of hypotheses and a "real world" perspective for their research, whether it be observational, theoretical, or experimental in nature.

In this chapter, we discuss the implications of using observation for learning ecology and increasing environmental awareness. Although we borrow from it, we do not intend to review the large body of literature in educational theory or practice (see for example, Brewer and Smith 2011; Hayes 2009; Bowen and Roth 2007) or "environmental psychology" (e.g., Clayton and Meyers 2009; Aanstoos 1998). Rather we use the salient findings from this research—which point to the importance of early exposure to nature, the value in any nature experiences, and the need to develop positive psychological feedback from nature exposure—as a framework to relate how our experiences as students, teachers, and parents have reinforced our own faith in the power of nature's classroom.

## Ecological Observations as a Developmental Tool

Children exposed to nature will quickly develop strong observational skills. Many of us have been in the field from a very young age (perhaps that is why we choose a career in ecology). Early contact with nature not only shaped our cognitive perception of nature, but also affected our connection to the natural world (Taylor and Kuo 2006). The flexibility of our brains in early childhood allows for rapid information acquisition and pattern-finding ability, thus setting the basal conditions to be able to construct synthetic perceptions of complex ecologies. Given a safe environment, young children set this process in motion by conducting a series of ever-more-elaborate observational experiments with the world around them (Gopnik 2009), learning from both successes and failures. For example, Aníbal's son was three years old when, one day at the grocery store, the boy grabbed a leaf from a decorative shrub, smashed it into his hand and smelled it, but perceived no smell. His father had to explain that this was a plastic plant. At this young age, he was already familiar with the fact that plants have particular smells, a secret that his father taught him in his own backyard. Although the immediate outcome was bewildering, the experience reinforced and refined the lesson that careful observation can be used to make discoveries and find order within a world full of stimuli.

Direct observation of nature, if accompanied with enjoyable experiences, leaves a lasting impression about the valuation of nature. Children whose relationship to nature started with exciting outdoor activities will be much more likely to develop a strong attachment to nature (Clayton and Myers 2009). For example, Nadia Lalak found that by bringing young children into forests in Australia and combining the field experience with a magical storytelling component, the children will quickly connect to nature. She reports that young children, aided by the dynamics of storytelling, use all their senses to experience the environment that surrounds them. Children quickly become "engaged through their senses into heightened awareness" in an otherwise unfamiliar environment (Burns 2005, p. 72; Lalak 2003). Early exposure to nature has important psychological benefits that go far beyond the appreciation of nature. For example, self-esteem, self-confidence and social behavior in children have been shown to improve after wilderness experiences (Taylor and Kuo 2006).

If productive success in ecology may stem from a personal, sensory experience, a corollary may be that divorcing natural history and observation may not just be detrimental to ecology as a science, but to individuals on a personal level as well. "Screen time" is replacing "forest hours," even for ecologists, and even our best screens are still massively lower in resolution than nature. Regrettably, we are also urbanizing ourselves, making nature a more and more a separate entity that we visit once a week if we're lucky or maybe only once a year for our summer vacations. In the United States, dire warnings about abductions blasted on cable news and even legal restrictions on letting children roam their neighborhoods and countryside without supervision are putting more barriers between kids and nature. It is hard to imagine a Geerat Vermeij or Daniel Kish (see Chapter 3) developing such keen observational senses in a world where we don't even let our sighted kids out of our sight.

Some have suggested that the separation of humans from nature has decreased the sensitivity of our observation skills as our cognitive and sensorial systems are now trained only under stressful, "sharp-elbowed" surroundings of urban environments (Berman, Jonides, and Kaplan 2008), which may help us avoid cars, but may overwhelm our abilities to detect more subtle changes in nature. The disconnection to nature has also been linked to higher rates of obesity, and potentially higher propensity toward mental illnesses such as Attention Deficit Hyperactivity Disorder (Louv 2005; Taylor, Kuo, and Sullivan 2001). At the same time, we are becoming handicapped in our ability to observe nature as our senses atrophy due to the lack of use. This disability can be direct and quantifiable, as in an epidemiological study revealing that children who spend more time outdoors have lower rates of myopia (Gwiazda and Deng 2009), or it may appear in the more qualitative sense that becoming a good observer requires immense practice, which is forfeited when we grow up indoors.

However, many people have rebelled against this trend toward the indoors, finding new ways to get back to nature. There is even a movement called "free-range kids" to encourage parents to let their kids explore the world on their own. The movement was started by a mother who wrote a newspaper column on why she let her nine-year-old son ride the New York City subway alone, only to find herself just days later on national television being called "the worst mother in America" by hyper-safety-conscious parenting "experts" (freerangekids.wordpress.com). More-formalized programs, sometimes primarily aimed at dealing with public-health concerns about a sedentary lifestyle, are encouraging more outdoor and nature activity for children. For example, No Child Left Indoors, a project run by government and nongovernment organizations in the American state of Connecticut, aims to "introduce children to the wonder of nature—for their own health and well-being, for the future of environmental conservation, and for the preservation of the beauty."

## Ecological Observations as an Elementary Learning Tool

An emerging lesson from these programs is that the nature experiences we try to foster need not be perfect forays into "ecotopias"—special trips to the Grand Canyon or the Great Barrier Reef. If we look to experience and share only idealized "pristine" natural areas, we may miss the chance to observe the nature that inhabits our cities, parks, and suburban environments. Biodiversity and ecosystem processes can be observed next to home (Brewer 2002). In many cases, the novel ecosystems (*sensu* Hobbs et al. 2006) that surround us, and which are under the strong influence of humans, are going to be the same ecosystems where we will spend most of our lives and where biodiversity should be conserved (see, for example, Pauchard et al. 2006). Observing and understanding these novel ecosystems, which can be dominated by non-native species, should be promoted regardless of their origins or level of preservation. Brendon Larson reminds us in Box 9.1 that even anthropogenic systems provide a unique opportunity to introduce the complexities, and wonder, of ecological systems.

Observation of nature can also help us avoid the frustration caused by the urgency of environmental catastrophes. As David Sobel (1996) suggests, children who are faced with environmental negativism can quickly become disenchanted with nature conservation, considering it a lost cause. Observation of ecological systems, even in heavily impacted environments, can generate hope out of a situation of despair for children and adults. For example, in Los Angeles, California, even small "pocket parks" at the dead end of streets that abut the concrete flood-control channel once known as the L.A. River, as well as the few remaining patches of natural streambed in the river, are becoming sources of ecological study, nature recreation, and neighborhood pride (J. Linton, pers. comm. to RDS). Knowledge and enthusiasm gained from these small, hopeful experiences can be contagious between peers, and even with other adults, which can have important repercussions for environmental awareness (Clayton and Myers 2009). Parents and teachers need to understand that nature is everywhere, even on an apartment balcony, and that children can be encouraged to embrace this nature as early as possible.

For students in their adolescent years, learning ecology through the observation of nature can have an important influence in shaping their personalities and fostering their enthusiasm about careers in ecology and natural-resource conservation (Thomashow 2002). Even if this is not their priority, learning through observation of their environment

BOX 9.1

**Observing Invasive Species and Novel Ecosystems in Urban Areas**

BRENDON LARSON

I grew up in the countryside and my life as a naturalist derived from long hours spent rambling along a local creek, looking for new plant species, learning bird calls, and trying to catch that one big dragonfly that was always just out of reach of my net.

Unfortunately, such childhood experiences are increasingly uncommon. The majority of the world's people now live in urban areas—and the percentage is increasing. The future of earth's biodiversity will depend on the relationship—or lack thereof—that these people develop with nature. While some of them will visit "wilderness" areas, which themselves face greater and greater threats due to the global scale of our impacts, most of them won't. Thus, their relationship with nature will mostly come about through observations of nature around them—in their backyards, local parks, and schoolyards, perhaps supplemented by a few transformative experiences outside of their urban homes.

Accordingly, ecologists need to reflect upon their own tendency to emphasize the importance of rare and endemic species found in protected wilderness areas that are too often distant from urban areas.

It is particularly useful to reflect on invasive species in this context. Urban environments are largely disturbed, which contributes to a preponderance of introduced and often invasive species.

will help them to have a broader understanding of the world they live in and the consequences of their actions for the biosphere (Kaplan and Kaplan 2002). Nonetheless, teenagers usually feel alienated from nature, and short-term experiences in nature do not seem to change this notion. For example, Haluza-Delay (2001) reports that after a 12-day field trip teenagers felt that nature only exists in more pristine environments and that there is little they can do to conserve it. One way to break through this perceived barrier is to involve adolescents in a working project with direct consequences for their environment. For example, when teenagers help to survey and plan a protected area they can see the direct effects of

It concerns me that so many environmental education programs now emphasize the "horrors" of invasive species, when these are the species that comprise the natural environment of so many children growing up in urban areas. Often, they are taught such lessons at what I consider too young an age, when the emphasis should instead be on becoming comfortable, exploring, and playing in this environment.

While there are ecological lessons and distinctions here, I feel quite certain that the risks of educating students with the idea that these landscapes are tainted and unworthy is untenable. The students might not want to go back. Instead, show them plants and teach them their stories, maybe even highlighting how these biological communities are now as cosmopolitan as contemporary cities. Consider the ecological functions that these communities might serve.

This might even be a first lesson in scientific "objectivity," in teaching them to observe the world around them before making judgments about its quality—perhaps even learning how difficult it is to justify distinctions between "good" and "poor" quality. The jump to higher-level concepts related to biogeography, dispersal, and biodiversity can come later.

Such lessons might help to nurture these urban children's urge to explore, much as I did, so that they grow closer to the nature that exists around them and perhaps even try to catch that one special living being that is just out of their reach.

their actions (Thomashow 2002). Here we see a clear convergence between educational, psycho-social, and scientific outcomes, because this kind of "project-based" or "problem-based" learning is exactly what education advocates are pushing for within and beyond the sciences (Darling-Hammond 2008).

Unfortunately, the educational system for elementary and high school students is extremely classroom oriented, and the opportunities for outdoor educational activities compete with other educational needs. Our traditional educational systems are based on closed environments with occasional outside "breaks" that are not considered part of the curricula.

Certainly, most elementary education is not integrated directly with the natural world, and sometimes outdoor time is foregone completely. In part, this may be explained by a lack of adequate training for school teachers. In addition, there is a notion that outdoor activities are expensive and require larger organizational efforts (Brewer 2002). Finally, teachers at all levels report to us that the pressures to keep students inside has also increased in recent years as requirements for standardized testing have become overwhelming and fears about uncontrolled hazards in the outdoors have gripped school administrators.

Yet ecologists and educators have designed several methods to increase children's connections to nature, combining observation and experimentation with the latest techniques for effective learning (Knapp and D'Avanzo 2010). By no means do all such activities need to be formalized. Non-curricular "informal science" activities such as creating native plant gardens or organic orchards or environmental filmmaking may also help to inspire a closer connection to nature in young students while providing improved learning outcomes overall (O'Neill 2005; Rennie et al. 2003). Carol Brewer, a recipient of the Eugene P. Odum Award for Ecological Education from the Ecological Society of America, has advocated for the use of schoolyards as outdoor biology classrooms (Brewer 2002). Based on experiences in North and South America, she has shown that areas close to schools, even in very disturbed environments, can serve as nature labs for students of all ages. Interestingly, most teachers initially admit they have difficulties in endorsing this idea, because "ecology happened on field trips" (Brewer 2002). However, given the appropriate conditions, teachers soon realize the advantages of using their own schoolyards for teaching ecology. First, it is much more affordable and less logistically complex than organizing field trips. Second, they can use it continuously over the course of a year. Third, and probably most important, it gives the students a sense of ecological place.

There are increasing opportunities for ecological education for all age groups. Many of these opportunities fit the same citizen-science model we discussed early, but provide platforms and incentives that are age and developmentally appropriate. For example, Rafe's daughter at age ten became enamored with the "Project Noah" website (projectnoah .org), which encourages young and old explorers to photograph natural organisms and post them to an account on the website. Other members of the site can then help identify the organism and add their comments. Achievement "patches" are awarded for reaching different goals, such as making 15 observations of amphibians, or participating in "missions" focused on particular taxa. Children who use the site are incentivized to get outside and observe, and by exploring the observations of other users throughout the world, and by considering the feedback they get on their own observations, they begin to learn the basics of classification, biogeography, and ecology. Such a technology-enabled observational program creates opportunities for the youngest generation of "digital natives," but it also raises questions about the appropriateness of technological intermediaries between people and nature. Teachers who are starting to integrate nature observation and technology are discovering firsthand the great promise and many challenges in this precarious interface, as Kristin Wisneski and Barron Orr share with us in Box 9.2.

### Ecological Observations in Higher Education

Values associated with nature are usually underrepresented in the aspirations of young students when applying to college, especially in developing countries. This problem may, in the long run, undermine the development of disciplines associated with ecology and the conservation of nature. For example, in Chile, programs associated with nature conservation have proliferated in universities in the last two decades. However, they have not been able to attract students as expected. This may be because environmental awareness has only recently become widespread in Chilean society, but also because students fear that the job market is not strong in conservation and life sciences–oriented career tracks.

Ecology education programs in schools may help to reverse this trend. In Chile, EXPLORA, a government-sponsored program aimed to bring science from the universities to high schools, has been particularly successful in connecting environmental issues to their scientific underpinnings and attracting students to nature-related careers. Many students

BOX 9.2

## Taking *Akshen* in Communities and the Environment with Mobile, Social, and Geospatial Technologies

KRISTIN D. WISNESKI AND BARRON J. ORR

Every moment spent outdoors offers a multitude of learning opportunities for youth and adults. Each opportunity begins with an observation. The things we see, smell, touch, taste, and hear are captured in our memories and ignite a reaction that has the potential to inspire and excite, stimulating our interest and imagination, creating the opportunity for discovery and learning. A new era of interaction with one another and our environment has arrived. Some say that firsthand experience of nature, especially by youth, should be untainted by the distraction of electronic devices. We have learned there are differences in devices and the levels of distraction they create when technology enters the outdoor classroom. A global positioning system (GPS) receiver that primarily delivers location information offers far fewer distractions than a GPS-enabled smartphone that can simultaneously provide access to the Internet and music, as well as multiple channels of communication and social networking. Electronic or not, distractions have always existed in the same space where learning takes place, and that space is increasingly being filled with ubiquitous access to—everything. Even as we find ways to create opportunities for youth to experience nature outside their digital environments, surely it behooves us as educators to explore ways of creating opportunities for observation and inquiry within the technological space where so many youth want to spend their time creating and sharing information.

The challenge: youth are spending more time connected to the digital world while interest in science, technology, engineering, and math (STEM) declines. Is it possible that the same technologies that are used to enhance learning at school also mean increased "screen time" and therefore decreased physical activity and time spent outdoors in nature? Can we turn this problem into an opportunity with the rapid growth of mobile communication and location technologies? As part of an interdisciplinary team at the University of Arizona, we set out to understand where youth engagement, technology, and learning intersect. With a youth-driven, participatory, and formative research and development process we designed and implemented smartphone applications (apps) for youth to collect observations for problem solving that would contribute to a collaborative group database stored online in a social network called *Akshen*. The Akshen apps and website drew on our experiences from the past using GPS receivers, web-based mapping tools, and the scientific method. These experiences, combined with contemporary and emerging educational theories and practices like problem-based learning (Bransford 2000; Hmelo-Silver 2004; Darling-Hammond 2008), contributed to a framework for educators to help youth identify and solve problems in their communities and the environment. Youth who wonder why the wash or stream behind their house always has trash in it can create a team of friends to investigate and analyze the problem on a community-wide scale and then create a strategy to share information and mobilize community members to help discover solutions to the problem and create change. An afterschool youth group that lacks a physical meeting space can map out and analyze the safe and dangerous places in their community to determine ideal locations for a new youth center and bring their findings to city council.

Unlike traditional classrooms, the outdoor classroom aided by technology lacks physical and mental boundaries. Learning can happen anytime and anywhere. Youth enjoy making "posts" and "status updates" as they explore their community and environment. By placing these familiar and enjoyable experiences within the context of science-related problem solving, youth are informally learning new skills while developing knowledge and a greater understanding of their surroundings. In the process, they become aware of the community and environmental challenges around them, and then learn how to pose good questions and collect the information necessary to answer those questions on the path toward finding solutions. Perhaps in the near future, "outside" will not only be the space between home and school, but the place where observations become the norm, making the outdoors less foreign and more a place to play, learn, and connect physically—and digitally—with the living world.

who went through this program were later drawn to study science-related majors in their college years.

But the hands-on experiences that programs such as EXPLORA provide need to be present in the undergraduate environment in order to maintain the enthusiasm that they generate. At the higher-education level, field observation should be a natural part of learning and doing ecology, but unfortunately, students are being exposed to fewer field opportunities. This trend is being discovered with alarm in other higher-education fields, such as political science (Schwartz-Shea 2010), which has gone through its own retreat into reductionism and theory at the expense of scientists who understand political systems in a broad and comparative way through a background in field observation.

For students majoring in natural resources and related disciplines (e.g., agronomy, forestry) there is a good chance they will have opportunities to interact with and directly observe ecosystems in different states of human transformation. But not all programs, not all courses, and not all instructors pay the same attention to the importance of observing nature. Nonetheless, as with early nature education, we should not let the search for ideal observational experiences cause us to forego valuable opportunities that are readily available. Even short field activities in university "backyards" during the semester can have disproportionate value. For example, Jake Weltzin, director of the U.S. National Phenology Network, set up a short hour-long field trip for one of Rafe's courses to measure phenological stages (budding, flowering, fruiting, etc.) in a small desert native-plant garden on the University of Arizona campus. Although the experience was brief, students reported being much more aware of phenological phases in other plants in their desert environment in subsequent weeks. Likewise, a simple but informative exercise in an introductory ecology class is to divide the students in the field into small groups and ask them to record all the "elements" and "processes" they can observe with all their senses. When the groups reunite, those groups that worked in similar habitats may find that they have completely different "narratives" of what they have observed. By transmitting these narratives to their classmates, they come to understand the value of having multiple observers, and more important, they realize that they can also "create" ecological knowledge.

As with younger children, undergraduates' experiences in observation and ecology are more strongly influential when combined with satisfactory personal psychological experiences. For almost ten years, Aníbal has been taking forestry students to one of the most marvelous national parks in Chile, Conguillio National Park, an intricate mosaic of volcanoes and monkey-puzzle trees (*Araucaria araucana*). The field day includes a hike to the treeline and multiple stations for the observation and discussion of ecology and protected-area management. Although not all the students are particularly focused on making careers of ecology and conservation, the shared feelings after this long day hike are amazement at the natural world and a sense of personal accomplishment. For some students, this first exposure to the elevational changes in vegetation and the beginning of the alpine zone, occurring amidst gorgeous surroundings and combined at different times during the experience with physical effort, solitude, team spirit, surprise, and achievement, leaves a deep, long-lasting impression.

Of course, longer field courses bring a unique experience to undergraduate students. For example, field-sampling techniques and field-observation techniques are usually only learned over long periods in the field that are, unfortunately, unavailable during the regular semester, and such longer courses conducted during breaks are increasingly rare due to budget cuts, safety and legal concerns, and competing priorities for students' time. But between brief mind-opening exercises and extended field-learning opportunities, active student participation in nature observation can be maintained, for example, by making a class requirement to keep a detailed field diary, which can comprise open-ended observations or responses to particular questions geared to the local environment and season.

Graduate school poses a new set of challenges for observations in ecology. When a recently enrolled PhD student starts out his research, there are basically two ways to figure out his dissertation theme. He may get a structured assignment from his adviser based on current grants, or a chance to "wander around" through different topics and approaches (Nunez and Crutsinger 2008) and ultimately settle on a topic that is (with a little luck) both interesting and likely to be successful in the time frame of a dissertation. Unfortunately, pressure builds up to hurry up this process and in many cases students know very little about their study system, but are pressured toward studies that have higher chances of publication.

This is not trivial. Accreditation processes, at least where we work in Latin America and the United States, increasingly emphasize the importance of reducing time until graduation. In addition, increasing competition among PhDs requires a larger number of publications to ensure a post-doc or faculty position. Are these two factors unequivocally conspiring against observational methods in ecology? It depends. On first consideration, it would seem that observational approaches necessarily take longer than other, more reductionist approaches, but this does not have to be the case. The smart use of existing data sets may help to compensate for the uncertainties brought by observational approaches. Even if the PhD dissertation is mainly experimental, careful observations of the study system provide essential contextual frameworks for the research. Observations may facilitate hypothesis development, help to focus data collection efforts, and even reduce the risk of failed experiments.

Students are so eager to start generating results that can contribute to their dissertation goals that they may forget the need to understand the overall system they are studying. Professors often complain that students know a lot about their own work, but outside their particular experiment or field samples they demonstrate little general knowledge. Even a brief pause to attend to observations of a system may be beneficial. For instance, if a student is interested in studying plant-plant interactions in a specific environment—a project amenable to experimental manipulations—a short field season of recording the abundance, phenology, and spatial aggregation of the plants that constitute the model system may prove to be mind-opening and may even change the expected course of the research. Unfortunately, many students rely only on the literature and may be missing the opportunity to investigate novel hypotheses that arise only by observing the plants and their interactions in the field. The role of professors here may be the same as the role of responsible parents in the twenty-first century—to push would-be ecological observers out of the confinement of the lab or the home and into the dynamic world around us.

Higher-education students can now stock their growing field knowledge back into the earlier reaches of the learning stream through programs that give them the opportunity to teach ecological lessons to primary and secondary students. The GK-12 program, initiated by the National Science Foundation in the United States in 1999, brings graduate students into the classrooms of elementary and high schools (McBride et al. 2011). At the University of Montana, this program has included the promotion of schoolyards as outdoor laboratories to which graduate students bring their own skills and interests—from soil ecology to plant phenology to connecting young scientists with citizen-science efforts such as Project Budburst (budburst.org). The benefits of this exchange permeate in both directions. Graduate students improve their teaching skills and children and adolescents get exposed to both the enthusiasm of young scientists and the firsthand knowledge of researchers at the leading edge of research in ecology.

### The Natural Draw of Observational Education

In this part of the book, we have discussed how ecology and the observation of nature can be integrated with many societal endeavors. As observation plays a key role in forming new directions in ecological science, and as this science is being shared more widely and with more consequence than before (see Chapter 8), now is the time to ensure that education at all levels trains new generations of citizens how to observe nature and how to understand those observations in the context of a dynamic social and ecological world.

Fortunately, there is an innate drive toward developing this kind of learning, which may only be superficially masked by the urban lifestyles and screen devotion that separate many of us from nature. This underlying drive is often revealed in the subtle interactions that many field ecologists have had while doing fieldwork in a publicly accessible place, where inevitably bystanders make wistful comments such as, "I wish I could do what you do. You get paid to go to places where I can only go on vacations." Behind the naïve impressions of these inquisitors (they're probably not thinking about the frustrating equipment failures, the relentless mosquitoes, or the endless process of writing and rewriting grants in order to get out into the field in the first place), there is a basic fact—as humans, we like to be in touch with nature, we like to observe nature, and we like to understand nature. Otherwise we could have not evolved into the species we are now.

Sometimes we overlook this simple and logical connection, and we might even be alienating people from ecology by portraying this discipline as a complex, abstract endeavor. At the same time, in a scientific sense, observation will not be useful unless we connect it to conceptual constructs that can be widely appreciated. The intuitive nature of observation-based ecology gives it a natural foundation for these constructs, while it also provides a source of tangible and readily accessible material with which to build connections between unique personal experiences (such as those generated in a field course or on a nature walk with field notebook and binoculars in hand) and advanced scientific knowledge. We conclude this book in the following pages with a glimpse of what we think the future architecture and architects of these societally integrated ecological constructs will look like.

# Conclusions

Science is often portrayed as an incremental process of building an edifice of knowledge, brick by brick. The benefits of this approach are said to accrue from the precision and rigor that comes from studying a carefully chosen set of variables at small spatial scales. Robert Paine argues that the understanding of ecological systems at large scales can be built this way, noting, "Even the smallest bricks, if solid enough, can be used to construct the largest building." But this analogy and, by extension, this way of doing science, which served us adequately in the twentieth century, doesn't hold up when we try to make sense of rapidly changing ecological systems that are increasingly intertwined with complex human behaviors. The problem is no longer how large we can make the building, but rather how quickly it can be made, and even whether a building is really what we need to bring together the growing body of scientific understanding of the world. The brick-by-brick approach would be fine if we had limitless time to build ecological understanding, but it is not scaled to the dimensions of time in which we need answers right now. The foundation of ecology—the natural world and its networked relationships—is collapsing faster than bricklayers can build an understanding of it.

Moreover, bricks don't accurately reflect our current abilities or technologies. The structure of ecological science is now being built out of materials that are far more resilient and flexible, which allows us to quickly make large leaps in the construction of our understanding without sacrificing the strength of our inferences. Finally, the brick-by-brick analogy assumes that what we want, and that the best we can aspire to, is a structure which is merely an enlargement of our small-scale view. A brick structure is linear and predictable, but the world we live in now is neither.

### A Changing Landscape for Ecology

The accelerating change in ecological systems requires even more cohesion in ecological knowledge. Early predictions of how ecological systems would respond to climate warming assumed relatively linear changes—basically, species would march across lines of latitude toward the poles to stay within their physiologically optimal environments as temperatures warmed. But as these models became evermore complex, taking into account basic biological facts such as how shrub and tree species have different constraints on how far they can relocate their ranges from year to year, scientists soon began talking about the future bringing "no-analog" communities (Williams and Jackson 2007). That is, we can't make a simple analogy that a birch forest of the future will be just like today's birch forest, only shifted 500 km northward, because the components of that forest will actually change their populations and sizes and physiology at different rates from one another. With increasing scenarios of biotic exchange and anthropogenic disturbances, we face the appearance of novel ecosystems (Hobbs et al. 2006), where our current ecological understanding may fail to predict new interactions, and our current methods of ecological restoration and management may not apply (Seastedt, Hobbs, and Suding 2008).

We have really already entered the era of no-analog ecology. There are multiple large-scale ecological changes that have few or no analogies in Earth's history. An enormous trash gyre swirling in the Pacific. Entire bodies of water with no living thing in them. Nutrient concentrations higher than ever experienced before. Ecological science that assumes the past can be replicated infinitely into the future simply can't survive in this no-analog world.

In a no-analog world, concepts developed in isolated small-scale environments will be increasingly difficult to generalize. At the same time, observational historical data taken over short time scales, from single sites, or on a small number of variables are only useful for telling us that change has occurred, but not how it occurred or whether similar change is likely in the future. Water managers for years thought that their century or so of records from channel gauges were enough to establish the rules of floodplain management (Craig 2010). They used measures like the "100-year flood" to plan for worst-case scenarios as they engineered hardened solutions, like levee walls, to protect cities or maintain water supplies to agriculture. In our no-analog world, however, not only is the 100-year flood an inadequate measure of the variation in water flows we see today, but a whole host of ecological changes, such as new invasive species, altered fire regimes, and changing patterns of human development are changing the way any level of water flow will affect human and ecological systems (Betancourt in press).

To deal with these multiple levels of change with multiple causal drivers that are themselves changing, ecology needs to become a more adaptable science, and observation can be a catalyst for this adaptation. One hallmark of adaptable systems is an intensive ability to sense conditions and variation in the world. This will require ongoing and expanded commitment to monitoring ecological systems in the future, and, ironically, greater attention to the past as well. Ecologist Julio Betancourt has written, "In the face of a nonstationary world, history is alive and well and historical ecology is more priceless than ever" (Betancourt in press). Betancourt's point is that perhaps the only way we can see what is coming is by more fully exploring conditions of genetic structure, populations, and communities in relation to past climates and conditions so that we can isolate those modern factors that cause diversions from the historical patterns. Getting to this point, where ecologists can adaptively learn from the past, will require networks of observers committed to re-examining historical records with modern technology and knowledge, combining multiple historical records, and continuing record keeping into the future.

But just as organisms must both observe change in the world and change themselves in order to be adaptable, ecology must use its expanding observational capacities to facilitate its own change. In this book we have argued that ecology is already being pushed, and is pulling itself, into new niches that will make it more adaptable and useful to the challenges of this century. In a way, this transformation is just an accelerated version of the usual metamorphoses of science. Different fields of science and even sub-branches of fields like ecology are always in flux with regard to dominant approaches and methodologies. Through the years, ecology as a whole has shifted from a science of observation and discovery to one of experimentation and theory and now to a science in which observations can be used multi-modally—building up an inductive understanding, or deducing mechanisms in a strong-inference framework. What we see now is an ecology that has grown recursively from all of these early stages into a more integrated science that will, overall, trend toward greater integration in the future. E. O. Wilson was prescient about these integrations in his 1998 book *Consilience* (Wilson 1998), which was received with some skepticism by the scientific community, but in a very short time many of the visions he laid out have come to pass.

Even as ecology flows toward greater use of observation and as it braids out to meet other disciplines, there will be important back-eddies that enrich the science in their own way. Some aspects of observational ecology, especially its newest branches, will most likely turn into manipulative sciences. For example, environmental genomics is still just stretching out from work with "model systems" in the laboratory to a descriptive phase in which patterns of gene expression are correlated with the varying environmental conditions observed where the specimens were collected (Eckert et al. 2010). But as technology and our understanding of genomics increases (and the cost in time and money for conducting genomics research continues to plummet), undoubtedly a heavy focus on manipulating variables and identifying genome-wide responses will eclipse the exploratory mode of today (A. Hancock, pers. comm. to RDS, Feb. 2011).

In other cases, old speculative ideas are reemerging as exploratory and descriptive sciences with impressive new powers. In the early twentieth century both Warder Allee and Edward Ricketts made forays into what we now call network science. Allee felt that a kind of adaptability was conferred to populations that were "subdivided into many small local populations almost but not quite completely isolated from each other"

(Allee 1938). Ricketts in his notes envisioned "an exact and a quantitative science in which the vectors representing [ecological] relationships, their direction, extension, and strength or intensity, would be considered and evaluated" (Tamm 2004). These quotations encapsulate both the descriptive and analytical aspects of modern network science, but their authors probably couldn't have fully realized its role today as a multi-functional observational tool. Network science has produced a set of theories about dynamic relationships that can be tested with experiments on everything from social insects to food webs to the relationships of movie stars (Barabasi 2003). Network science connects pattern to process, as in the observations of the changing relationships of the 9/11 terrorists, which provided deep insight into the mechanisms of their operation (Sageman 2004; Krebs 2002). It also provides a map and strategies for improving communication and productivity among scientists, as we discussed in Chapter 6. And it can be used as an analytical tool that works universally across disciplines for making sense of a complex world. Ecologist Eric Berlow, for example, gave a brilliant three-minute TED (Technology, Entertainment, and Design) talk in Oxford in 2010 that used the same network analysis he applies to Sierra Nevada lake food webs to transform a complicated government-created diagram of U.S. strategy in Afghanistan (which had been universally ridiculed in the media) into a very small set of truly actionable tasks (Berlow 2010). The ability to turn ideas that were once only conceptual into practical tools applicable across disciplines is one reason that twenty-first-century ecology is so exciting and so powerful.

## Integrating the Human Element of Ecology

There is now a growing and increasingly irresistible gravitational pull between the changes occurring to ecological systems and the way we study those changes. In almost all cases, humans are generating that force. The impacts of humans on ecological systems have been abundantly well documented, from local to global scales. But for a long time, ecologists either ignored humans, or at best treated them as a black box, a mysterious "dark force" that had an effect on the systems we were studying, to be sure, but one we didn't really want to delve into too much. Ecologists rarely incorporate human behavior directly into their experiments or observations, as if human behavior were just something that happens to a system which has been studied in detail, rather than an integral part of that system's ecological dynamics. Ivar and Iver Mysterud, refer to this as "research escapism" (Mysterud and Mysterud 1994) with potentially detrimental consequences, both in academic terms of making ecology appear to be far more narrow a science than it should be, and in real terms of limiting how well we understand ecological systems as a whole.

We now have the ability, and many would say the responsibility, to become more active in how we deal with human impacts on ecological systems. This doesn't mean just becoming a scientific advocate for environmental protection, although that is still sorely needed 50 years after Rachel Carson stuck her frail neck into the political fray with the publication of *Silent Spring*. It does mean integrating people and their behaviors into our hypotheses, our models, our experiments, and our observations. Just as physicists are now boldly examining the very nature of the dark forces and dark energies that once just muddled their calculations, we too can get down to the first principles of human behaviors and their effects on ecology. Fortunately, we don't need multibillion-dollar particle accelerators to explore our dark forces. We just have to bring them to light by talking with people and working with them and studying their evolution, exactly as we've been doing with all the other living things.

## Beyond Bricklaying: The New Architects and Architecture of Ecology

Who is and who will be driving this transformation of ecology? To use an ecological metaphor, ecologists for decades have debated whether the primary driving force in a particular ecological system was "top-down" (having to do with top predators controlling the dynamics of species below them) or "bottom-up" (being controlled by the amount of primary productivity in the system). The answer, not surprisingly, is that it's some of both, depending on where and when you study it (and by the way, all answers are subject to alteration due to climate change). To usefully capture this moment in history, ecological science itself will likewise need to rely on both top-down and bottom-up forces.

Top-down change is occurring in ecology, but much more slowly than bottom-up transformation, which seems to be happening automatically, as a natural adaptive response to our changing world. In the ecology of the science of ecology, the top-down forces tend to be faculty committees, editors, and review panels made up of mid-career and senior scientists who determine who gets jobs, journal publications, and research funding. But the primary productivity is ideas, and these can be equally owned by senior researchers or the newest ecologists. Indeed, today's students are embracing interdisciplinary observational methods and even teaching them to their advisors. These students will eventually develop and become tomorrow's top-down forces as they take roles as faculty chairs, foundation program officers, and editors, but this will take a long time.

Thus, it is incumbent upon today's leaders and gate keepers in ecology to reexamine the inherent assumptions and reflexive biases that we have all developed through time. A good place to start is by counting to ten every time we feel the urge to comment on a grant application or announce after a seminar, "Correlation does not imply causation," or "You can't infer process from pattern," or "Sounds like a fishing expedition." This pause for careful reconsideration might give us the opportunity to reflect more deeply upon the nature of the research itself, which is almost surely to be different from the type of research done when these phrases became clichés of twentieth-century ecological criticism. Then we can get to the questions that really matter in a practical sense for the future of ecology. Does the "sure thing" that a proposed set of carefully controlled small-scale experiments promises to yield make it inherently more fundable than a less-certain, more open-ended, but potentially more valuable observational study? Will a job candidate who has published eight good articles within a well-defined subdiscipline really be more valuable to the department than a candidate with a handful of articles scattered across disciplines and a couple of newspaper op-ed pieces and her own citizen-science program for K-12 students? There is no one correct answer to these questions, and that's exactly the point. In a changing world, anything that once sounded like a truism, a general law, or a "no-brainer" needs to be reexamined. We need to reevaluate the role of ecology in society, how much ecology is helping society, and how society values ecology. Observation may help us both to integrate society into ecology and to increase the value of ecology for society.

Over 100 years ago, Teddy Roosevelt derided the brick-by-brick kind of science he first encountered at Harvard, noting in a letter to George Bird Grinnell, "I know these scientists pretty well and their limitations are extraordinary, especially when they get to talking of science with a capital S. They do good work, but after all, it is only the best of them who are more than bricklayers, who laboriously get together bricks out of which other men must build houses; when they think they are architects they are simply a nuisance" (Cutright 1985). In his frustration with this mode of science he switched his career path from biology to politics, and ultimately he had a far greater impact on ecological systems and ecological science than if he had forced himself to follow the path of a bricklayer. He created the U.S. Forest Service and several major national parks and monuments. These not only protected nearly a million square kilometers of natural lands in his time, but created a legacy of ongoing landscape and biodiversity protection that has been emulated throughout the world. These lands then provided the spaces and questions on which many ecological studies have been conducted.

Roosevelt's personal transformation and the outcomes of that transformation were due to an idiosyncratic series of events that would have been entirely unpredictable. And yet it's no accident that the central player in this chaotic drama was an ecologist at heart. What Roosevelt brought to his political career came from his long hours in the field as an observer of nature—the ability to articulate connections across scales and fields of inquiry, and the intimate knowledge that energetic relationships rule all dynamics, whether they be between predators and prey or between political parties. These same skills are exactly what today's ecologists and the people we work with will need in order to create the kind of transformations necessary if we are to understand and protect ecological systems in this century.

There's no magic to developing these skills or using them. We've known how to do it since we were first human, intensely observing the world and recording those observations in stories, paintings, and in the ways in which we use natural resources. The problem now is that we have become so disconnected from the natural sources of our skills. This, in turn, starts a downward spiral, as described by Paul Dayton and Enric Sala (2001), in which the disconnection from nature leaves fewer people to care for and steward natural resources, and the degradation of nature that results leaves less people interested in or able to connect with it. We are at the point in the history of ecology and in the history of humanity's relationship to nature where we can, and must, turn this cycle of loss into a cycle of gain.

A renewed, more open, and more observational approach to ecology can be the catalyst for reversing the cycle. Opening our senses to changes and connections in the world, and teaching others how to open their senses, is just the first step. Fortunately, in the world we live in today, this process of discovery doesn't have to end with a collection of curious observations. We now have the holistic vision and the tools to see at once how each observation can connect to another, and how vast collections of observations—culled from all over the world and across long gaps in time—can help us understand a complex, interconnected, and relentlessly changing planet.

# REFERENCES


Aanstoos, Christopher M. "Humanistic Psychology and Ecopsychology." *The Humanistic Psychologist* 26, nos. 1–3 (1998): 3–4.

Aburto-Oropeza, O., E. Ezcurra, G. Danemann, V. Valdez, J. Murray, and E. Sala. "Mangroves in the Gulf of California Increase Fishery Yields." *Proceedings of the National Academy of Sciences of the United States of America* 105, no. 30 (2008): 10456–59.

Adler, Jonathan H. "Fables of the Cuyahoga: Reconstructing a History of Environmental Protection." *Fordham Environmental Law Journal* 14 (2003): 89.

Agrawal, A. A., D. D. Ackerly, F. Adler, A. E. Arnold, C. Caceres, D. Doak, E. Post, P. J. Hudson, J. Maron, K. A. Mooney, M. Power, D. Schemske, J. Stachowicz, S. Strauss, M. G. Turner, E. Werner. "Filling Key Gaps in Population and Community Ecology." *Frontiers in Ecology and the Environment* 5 (2007): 145–52.

Ainley, David G., Richard L. Veit, Sarah G. Allen, Larry B. Spear, and Peter Pyle. "Variations in Marine Bird Communities of the California Current 1986–1994." *California Cooperative Fisheries Reports* 36 (1995): 72–77.

Alexander, J. M., C. Kueffer, C. C. Daehler, P. J. Edwards, A. Pauchard, T. Seipel, and Miren Consortium. "Assembly of Nonnative Floras Along Elevational Gradients Explained by Directional Ecological Filtering." *Proceedings of the National Academy of Sciences of the United States of America* 108, no. 2 (2011): 656–61.

Allee, W. C. "Where Angels Fear to Tread—A Contribution from General Sociology to Human Ethics." *Science* 97, no. 2528 (1943): 517–25.

Allee, W. C. *Cooperation among Animals with Human Implications*. New York: Henry Schuman, 1951.

Allee, W. C. *The Social Life of Animals*. New York: W. W. Norton, 1938.

Allendorf, Fred W., Phillip R. England, Gordon Luikart, Peter A. Ritchie, and Nils Ryman. "Genetic Effects of Harvest on Wild Animal Populations." *Trends in Ecology & Evolution* 23, no. 6 (2008): 327–36.

Alter, S. Elizabeth, Eric Rynes, and Stephen R. Palumbi. "DNA Evidence for Historic Population Size and Past Ecosystem Impacts of Gray Whales." *Proceedings of the National Academy of Sciences* 104, no. 38 (2007): 15162–67.

Alvarez, Luis W., Walter Alvarez, Frank Asaro, and Helen V. Michel. "Extraterrestrial Cause for the Cretaceous-Tertiary Extinction—Experimental Results and Theoretical Interpretation." *Science* 208 (1980): 4448.

Alvarez, Walter, and Frank Asaro. "An Extraterrestrial Impact." *Scientific American*, October 1990, 78–84.

Applegate, Roger D. "Diversity and Natural History Observation in Ecology." *Oikos* 87, no. 3 (1999): 587–88.

Arias, P. A., R. Fu, C. D. Hoyos, W. H. Li, and L. M. Zhou. "Changes in Cloudiness over the Amazon Rainforests During the Last Two Decades: Diagnostic and Potential Causes." *Climate Dynamics* 37, no. 5–6 (2011): 1151–64.

Arnold, Stevan J. "Too Much Natural History, or Too Little?" *Animal Behavior* 65 (2003): 1065–68.

Assel, Raymond A., and Dale M. Robertson. "Changes in Winter Air Temperatures near Lake Michigan, 1851–1993, as Determined from Regional Lake-Ice Records." *Limnology and Oceanography* 40, no. 1 (1995): 165–76.

Attenborough, David. *Amazing Rare Things: The Art of Natural History in the Age of Discovery* New Haven: Yale Univ. Press, 2007.

Attum, O., B. Rabea, S. Osman, S. Habinan, S. M. Baha El Din, and B. Kingsbury. "Conserving and Studying Tortoises: A Local Community Visual-Tracking or Radio-Tracking Approach?" *Journal of Arid Environments* 72 (2008): 671–76.

Barabasi, Albert-Laszlo. *Linked*. London: Penguin Books Ltd., 2003.

Bart, David. "Integrating Local Ecological Knowledge and Manipulative Experiments to Find the Causes of Environmental Change." *Frontiers in Ecology and the Environment* 4, no. 10 (2006): 541–46.

Bartholomew, George A. "The Role of Natural History in Contemporary Biology." *BioScience* 36, no. 5 (1997): 324–29.

Bebber, D. P., M. A. Carine, J. R. I. Wood, A. H. Wortley, D. J. Harris, G. T. Prance, G. Davidse, J. Paige, T. D. Pennington, N. K. B. Robson, and R. W. Scotland. "Herbaria Are a Major Frontier for Species Discovery." *Proceedings of the National Academy of Sciences of the United States of America* 107, no. 51 (2010): 22169–71.

Berlow, Eric L. "How Complexity Leads to Simplicity." In *Technology, Entertainment and Design Conference*. Oxford: Oxford Univ. Press, 2010.

Berman, M. G., J. Jonides, and S. Kaplan. "The Cognitive Benefits of Interacting with Nature." *Psychological Science* 19, no. 12 (2008): 1207–12.

Beschta, R. L. "Reduced Cottonwood Recruitment Following Extirpation of Wolves in Yellowstone's Northern Range." *Ecology* 86, no. 2 (2005): 391–403.

Betancourt, Julio L. "Reflections on the Relevance of History in a Nonstationary World." In *Historical Environmental Variation in Conservation and Natural*

*Resource Management*, edited by John Wiens, Gregory D. Hayward, Hugh D. Stafford, and Catherine Giffen. New York: Wiley-Blackwell, forthcoming.

Beyers, D. W. "Causal Inference in Environmental Impact Studies." *Journal of the North American Benthological Society* 17, no. 3 (1998): 367–73.

Block, Barbara A, Steven L. H. Teo, Andreas Walli, Andre Boustany, Michael J. W. Stokesbury, Charles J. Farwell, Kevin C. Weng, Heidi Dewar, and Thomas D. Williams. "Electronic Tagging and Population Structure of Atlantic Bluefin Tuna." *Nature* 434 (2005): 1121–27.

Block, Barbara A. "Physiological Ecology in the 21st Century: Advancements in Biologging Science." *Integrative and Comparative Biology* 45 (2005): 305–20.

Bourlat, Sarah J., Thorhildur Juliusdottir, Christopher J. Lowe, Robert Freeman, Jochanan Aronowicz, Mark Kirschner, Eric S. Lander, Michael Thorndyke, Hiroaki Nakano, Andrea B. Kohn, Andreas Heyland, Leonid L. Moroz, Richard R. Copley, and Maximilian J. Telford. "Deuterostome Phylogeny Reveals Monophyletic Chordates and the New Phylum Xenoturbellida." *Nature* 444 (2006): 85–88.

Bowen, G. M., and W. M. Roth. "The Practice of Field Ecology: Insights for Science Education." *Research in Science Education* 37, no. 2 (2007): 171–87.

Bradbury, J. "Chernobyl: An Ecosystem Disaster?" *Frontiers in Ecology and the Environment* 5, no. 8 (2007): 401.

Bradley, Nina L., A. Carl Leopold, John Ross, and Wellington Huffaker. "Phenological Changes Reflect Climate Change in Wisconsin." *Proceedings of the National Academy of Sciences USA* 96 (1999): 9701–04.

Bransford, J. *How People Learn: Brain, Mind, Experience, and School*. Washington, DC: National Academies Press, 2000.

Breitbart, M., L. R. Thompson, C. A. Suttle, and M. B. Sullivan. "Exploring the Vast Diversity of Marine Viruses." *Oceanography* 20, no. 2 (2007): 135–39.

Brewer, C.A, and D. Smith, eds. *Vision and Change in Undergraduate Biology Education: A Call to Action. Final Report of a National Conference Organised by the AAAS, July 15–17 2009*. Washington DC, 2011.

Brewer, Carol. "Conservation Education Partnerships in Schoolyard Laboratories: A Call Back to Action." *Conservation Biology* 16, no. 3 (2002): 577–79.

Brook, R. K., and S. M. McLachlan. "On Using Expert-Based Science to 'Test' Local Ecological Knowledge." *Ecology and Society* 10, no. 2 (2005): r3.

Brooks, Michael. "Do You Speak Cuttlefish?" *New Scientist*, 26 April 2008, accessed online.

Brown, James H., and Arthur C. Gibson. *Biogeography*. St. Louis: C. V. Mosby, 1983.

Brown, J. *Macroecology*. Chicago, IL: Univ. of Chicago Press, 1995.

Buck, Carol. "Popper's Philosophy for Epidemiologists." *International Journal of Epidemiology* 4, no. 3 (1975): 159–72.

Burns, George W. 101 *Healing Stories for Kids and Teens: Using Metaphors in Therapy.* Hoboken, NJ: Wiley, 2005.

Camilli, R., C. M. Reddy, D. R. Yoerger, B. A. S. Van Mooy, M. V. Jakuba, J. C. Kinsey, C. P. McIntyre, S. P. Sylva, and J. V. Maloney. "Tracking Hydrocarbon Plume Transport and Biodegradation at Deepwater Horizon." *Science* 330, no. 6001 (2010): 201–04.

Canfield, Michael, ed. *Field Notes on Science and Nature*. Cambridge, MA: Harvard Univ. Press, 2011.

Cardinale, B. J, D. S. Srivastava, J. E. Duffy, J. P. Wright, A. L. Downing, M. Sankaran, and C. Jousseau. "Effects of Biodiversity on the Functioning of Trophic Groups and Ecosystems." *Nature* 443 (2006): 989–92.

Chariton, A. A., L. N. Court, D. M. Hartley, M. J. Colloff, and C. M. Hardy. "Ecological Assessment of Estuarine Sediments by Pyrosequencing Eukaryotic Ribosomal DNA." *Frontiers in Ecology and the Environment* 8, no. 5 (2010): 233–38.

Churchland, Patricia Smith. "How Do Neurons Know?" *Daedalus*, Winter 2004, 42–50.

Clark, James S. "Why Environmental Scientists Are Becoming Bayesians." *Ecology Letters* 8 (2005): 2–14.

Clarke, K. R., and R. M. Warwick. *Change in Marine Communities: An Approach to Statistical Analysis and Interpretation*. Second ed. Plymouth, UK: Primer-E, 2001.

Clayton, Susan D., and Gene Myers. *Conservation Psychology: Understanding and Promoting Human Care for Nature*. Chichester, UK / Hoboken, NJ: Wiley-Blackwell, 2009.

Cleland, Carol E. "Methodological and Epistemic Differences between Historical Science and Experimental Science." *Philosophy of Science* 69 (2002): 474–96.

Clements, Frederic E. "Nature and Structure of the Climax." *The Journal of Ecology* 24 (1936): 252–84.

Coltman, D. W. "Evolutionary Rebound from Selective Harvesting." *Trends in Ecology & Evolution* 23, no. 3 (2008): 117–18.

Coltman, D. W., P. O'Donoghue, J. T. Jorgenson, J. T. Hogg, C. Strobeck, and M. Festa-Bianchet. "Undesirable Evolutionary Consequences of Trophy Hunting." *Nature* 426, no. 6967 (2003): 655–58.

Corson, Trevor. *The Secret Life of Lobsters: How Fishermen and Scientists Are Unraveling the Mysteries of Our Favorite Crustacean*. New York: Harper Collins, 2004.

Craig, R. K. "'Stationarity Is Dead'—Long Live Transformation: Five Principles for Climate Change Adaptation Law." *Harvard Environmental Law Review* 34, no. 1 (2010): 9–73.

Cutright, Paul Russell. *Theodore Roosevelt: The Making of a Conservationist*. Urbana and Chicago, IL: Univ. of Illinois Press, 1985.

Cyr, D., S. Gauthier, Y. Bergeron, and C. Carcaillet. "Forest Management Is Driving the Eastern North American Boreal Forest Outside Its Natural Range of Variability." *Frontiers in Ecology and the Environment* 7, no. 10 (2009): 519–24.

Darling-Hammond, L. *Powerful Learning: What We Know About Teaching for Understanding*. San Francisco, CA: Jossey-Bass, 2008.

Darwin, Charles. *Voyage of the Beagle by Charles Darwin, with a New Introduction by David Quammen*. Washington, DC: National Geographic Society, 2004.

Davis, Daniel M. "Intrigue at the Immune Synapse." *Scientific American*, February 2006, 48–55.

Davis, Wade. "Last of Their Kind." *Scientific American*, September 2010.

Dayton, Paul K. "The Importance of the Natural Sciences to Conservation." *American Naturalist* 162, no. 1 (2003): 1–13.

Dayton, Paul K., and Enric Sala. "Natural History: The Sense of Wonder, Creativity, and Progress in Ecology." *Scientia Marina* 65, no. Suppl. 2 (2001): 199–206.

DellaSala, D.A. 2011. *Temperate and boreal rainforests of the world : ecology and conservation*. Washington, DC: Island Press. xvii, 295 p., 16 p. of plates pp.

Dellwo, Lisa M. "17 Years of Duke Forest Comes to a Close—Long-Term Ecological Experiment Will Continue to Provide Critical Data About the Impacts of Rising Carbon Dioxide." *Dukenviroment Magazine*, 2010.

Diamond, Jared, and James A. Robinson, eds. *Natural Experiments of History*. Cambridge, MA: Belknap Press of Harvard Univ. Press, 2010.

Diamond, Jared, and James A. Robinson. "All the World's a Lab." *New Scientist*, 27 March 2010, 28–31.

Diamond, Jared, and Ted J. Case. *Community Ecology*. New York: Harper and Row, 1986.

Dickson, B. G., E. Fleishman, D. S. Dobkin, and S. R. Hurteau. "Relationship between Avifaunal Occupancy and Riparian Vegetation in the Central Great Basin (Nevada, USA)." *Restoration Ecology* 17, no. 5 (2009): 722–30.

Dickson, B. G., S. Sesnie, E. Fleishman, and D. S. Dobkin. "Identification and Assessment of Habitat Quality for Conservation of Terrestrial Animals." In *Conservation Planning from the Bottom Up: A Practical Guide to Tools and Techniques for the Twenty-First Century*, edited by L. Craighead. Redlands, CA: ESRI Press, forthcoming.

Dorsey, J. H., P. M. Carter, S. Bergquist, and R. Sagarin. "Reduction of Fecal Indicator Bacteria (FIB) in the Ballona Wetlands Saltwater Marsh (Los Angeles County, California, USA) with Implications for Restoration Actions." *Water Research* 44, no. 15 (2010): 4630–42.

Eckert, A. J., A. D. Bower, S. C. Gonzalez-Martinez, J. L. Wegrzyn, G. Coop, and D. B. Neale. "Back to Nature: Ecological Genomics of Loblolly Pine (Pinus Taeda, Pinaceae)." *Molecular Ecology* 19, no. 17 (2010): 3789–805.

Ellison, Aaron M. "Bayesian Inference in Ecology." *Ecology Letters* 7 (2004): 509–20.

Elner, R. W., and R. L. Vadas. "Inference in Ecology—The Sea-Urchin Phenomenon in the Northwestern Atlantic." *American Naturalist* 136, no. 1 (1990): 108–25.

Farman, J. C., B. G. Gardiner, and J. D. Shanklin. "Large Losses of Total Ozone in Antarctica Reveal Seasonal Clox/Nox Interaction." *Nature* 315 (1985): 207–10.

Fleischner, Thomas L. "Natural History and the Deep Roots of Resource Management." *Natural Resources Journal* 45 (2005): 1–13.

Forman, Richard T. T. *Land Mosaics: The Ecology of Landscapes and Regions*. Cambridge, UK: Cambridge Univ. Press, 1995.

Francis, Robert C., and Steven R. Hare. "Decadal-Scale Regime Shifts in the Large Marine Ecosystems of the North-East Pacific: A Case for Historical Science." *Fisheries Oceanography* 3, no. 4 (1994): 279–91.

Freilich, J. E., J. M. Emlen, J. J. Duda, D. C. Freeman, and P. J. Cafaro. "Ecological Effects of Ranching: A Six-Point Critique." *BioScience* 53, no. 8 (2003): 759–65.

Fretwell, S. D. *Populations in a Seasonal Environment*. Princeton, NJ: Princeton Univ. Press, 1972.

Futuyma, Douglas J. "Wherefore and Whither the Naturalist?" *American Naturalist* 151, no. 1 (1998): 1–6.

Gardner, Janet, Peter Marsack, John Trueman, Brett Calcott, and Robert Heinsohn. "Story-Telling: An Essential Part of Science." *Trends in Ecology & Evolution* 22, no. 10 (2007): 510.

Garibaldi, Ann, and Nancy Turner. "Cultural Keystone Species: Implications for Ecological Conservation and Restoration." *Ecology and Society* 9, no. 3 (2004): 1. ecologyandsociety.org/vol9/iss3/art1/.

Garst, J., N. L. Kerr, S. E. Harris, and L. A. Sheppard. "Satisficing in Hypothesis Generation." *American Journal of Psychology* 115, no. 4 (2002): 475–500.

Gilchrist, Grant, and Mark Mallory. "Comparing Expert-Based Science with Local Ecological Knowledge: What Are We Afraid Of?" *Ecology and Society* 12, no. 1 (2007): R1.

Gilchrist, Grant, Mark Mallory, and Flemming Merkel. "Can Local Ecological Knowledge Contribute to Wildlife Management? Case Studies of Migratory Birds." *Ecology and Society* 10, no. 1 (2005): 20.

Gillis, Justin. " A Scientist, His Work, and a Climate Reckoning." *New York Times*, 21 December 2010.

Gladwell, Malcolm. "The Physical Genius." *The New Yorker*, 2 August 1999.

Gladwell, Malcolm. *Outliers*. New York: Little, Brown, 2008.

Gleason, H. A. "The Individualistic Concept of the Plant Association." *Bulletin of the Torrey Club* 53 (1926): 7–26.

Gopnik, Alison. *The Philosophical Baby: What Children's Minds Tell Us About Truth, Love, and the Meaning of Life*. New York: Farrar, Straus and Giroux, 2009.

Goyert, W., R. Sagarin, and J. Annala. "The Promise and Pitfalls of Marine Stewardship Council Certification: Maine Lobster as a Case Study." *Marine Policy* 34, no. 5 (2010): 1103–09.

Graham, Catherine H., Simon Ferrier, Falk Huettman, Craig Moritz, and Townsend A Peterson. "New Developments in Museum-Based Informatics and Applications in Biodiversity Analysis." *TRENDS in Ecology and Evolution* 19, no. 9 (2004).

Granek, E. F., and B. Ruttenberg. "The Protective Capacity of Mangroves During Tropical Storms: A Case Study from Wilma and Gamma in Belize." In *Western Society of Naturalists Annual Meeting*. Ventura, CA, 2007.

Greene, Harry W. "Organisms in Nature as a Central Focus for Biology." *Trends in Ecology & Evolution* 20, no. 1 (2005): 23–27.

Gremillet, D., R. H. E. Mullers, C. Moseley, L. Pichegru, J. C. Coetzee, P. S. Sabarros, C. D. van der Lingen, P. G. Ryan, A. Kato, and Y. Ropert-Coudert. "Seabirds, Fisheries, and Cameras." *Frontiers in Ecology and the Environment* 8, no. 8 (2010): 401–02.

Grinnell, Joseph. "The Role of the 'Accidental.'" *Auk* 34 (1922): 373–81.

Grobstein, Paul. "Revisiting Science in Culture: Science as Story Telling and Story Revising." *Journal of Research Practice* 1, no. 1 (2005): Article M1.

Groffman, P. M., C. Stylinski, M. C. Nisbet, C. M. Duarte, R. Jordan, A. Burgin, M. A. Previtali, and J. Coloso. "Restarting the Conversation: Challenges at the Interface between Ecology and Society." *Frontiers in Ecology and the Environment* 8, no. 6 (2010): 284–91.

Gwiazda, J, and L. Deng. "Hours Spent on Visual Activities Differ between Myopic and Non-Myopic Children." In *Myopia: Proceedings of the 12th International Conference*. Sydney Australia, 2009.

Hackett, E. J., J. N. Parker, D. Conz, D. Rhoten, and A. Parker. "Ecology Transformed: The National Center for Ecological Analysis and Synthesis and the Changing Patterns of Ecological Research." In *Scientific Collaboration on the Internet*, edited by G. M. Olson, A. Zimmerman, and N. Bos, 277–96. Boston: MIT Press, 2008.

Hall, L. S., P. R. Krausman, and M. L. Morrison. "The Habitat Concept and a Plea for Standard Terminology." *Wildlife Society Bulletin* 25, no. 1 (1997): 173–82.

Halpern, Benjamin S., Shaun Walbridge, Kimberly A. Selkoe, Carrie V. Kappel, Fiorenza Micheli, Caterina D'Agrosa, John F. Bruno, Kenneth S. Casey, Colin Ebert, Helen E. Fox, Rod Fujita, Dennis Heinemann, Hunter S. Lenihan, Elizabeth M. P. Madin, Matthew T. Perry, Elizabeth R. Selig, Mark Spalding, Robert Steneck, and Reg Watson. "A Global Map of Human Impact on Marine Ecosystems." *Science* 319 (2008): 948–52.

Haluza-Delay, Randolph. "Nothing Here to Care About: Participant Constructions of Nature Following a 12-Day Wilderness Program." *Journal of Environmental Education* 32, no. 4 (2001): 43–48.

Harrison, Freya. "Getting Started with Meta-Analysis." *Methods in Ecology and Evolution* 2, no. 1 (2010): 1–10.

Hartley, C. J., R. D. Newcomb, R. J. Russell, C. G. Yong, J. R. Stevens, D. K. Yeates, J. La Salle, and J. G. Oakeshott. "Amplification of DNA from Preserved Specimens Shows Blowflies Were Preadapted for the Rapid Evolution of Insecticide Resistance." *Proceedings of the National Academy of Sciences* 103, no. 23 (2006): 8757–62.

Hawken, Paul. *Blessed Unrest: How the Largest Movement in the World Came into Being and Why No One Saw It Coming*. New York: Viking, 2007.

Hayes, M. A. "Into the Field: Naturalistic Education and the Future of Conservation." *Conservation Biology* 23, no. 5 (2009): 1075–79.

Hazen, Robert. "Curve Fitting." *Science* 202, no. 4370 (1978): 823.

Helmuth, B. S. T. "Intertidal Mussel Microclimates: Predicting the Body Temperature of a Sessile Invertebrate." *Ecological Monographs* 68, no. 1 (1998): 51–74.

Helmuth, Brian, Joel G. Kingsolver, and Emily Carrington. "Biophysics, Physiological Ecology, and Climate Change: Does Mechanism Matter?" *Annual Review of Physiology* 67 (2005): 177–201.

Herrick, J. E., and J. Sarukhan. "A Strategy for Ecology in an Era of Globalization." *Frontiers in Ecology and the Environment* 5, no. 4 (2007): 172–81.

Herrick, J. E., V. C. Lessard, K. E. Spaeth, P. L. Shaver, R. S. Dayton, D. A. Pyke, L. Jolley, and J. J. Goebel. "National Ecosystem Assessments Supported by Scientific and Local Knowledge." *Frontiers in Ecology and the Environment* 8, no. 8 (2010): 403–08.

Hewitt, Judi E., Simon F. Thrush, Paul K Dayton, and Erik Bonsdorff. "The Effect of Spatial and Temporal Heterogeneity on the Design and Analysis of Empirical Studies of Scale-Dependent Systems." *American Naturalist* 169, no. 3 (2007): 398–408.

Hilborn, R., and D. Ludwig. "The Limits of Applied Ecological Research." *Ecological Applications* 3, no. 4 (1993): 550–52.

Hirzel, A. H., J. Hausser, D. Chessel, and N. Perrin. "Ecological-Niche Factor Analysis: How to Compute Habitat-Suitability Maps without Absence Data?" *Ecology* 83, no. 7 (2002): 2027–36.

Hmelo-Silver, C. "Problem-Based Learning: What and How Do Students Learn?" *Educational Psychology Review* 16, no. 3 (2004): 235–66.

Hobbs, R. J., S. Arico, J. Aronson, J. S. Baron, P. Bridgewater, V. A. Cramer, P. R. Epstein, J. J. Ewel, C. A. Klink, A. E. Lugo, D. Norton, D. Ojima, D. M. Richardson, E. W. Sanderson, F. Valladares, M. Vila, R. Zamora, and M. Zobel. "Novel Ecosystems: Theoretical and Management Aspects of the New Ecological World Order." *Global Ecology and Biogeography* 15, no. 1 (2006): 1–7.

Hoffmann, J. A. *Flora silvestre de Chile, zona central.* Santiago de Chile: Talleres Empresa El Mercurio, 1989.

Hofmann, G. E., and S. P. Place. "Genomics-Enabled Research in Marine Ecology: Challenges, Risks, and Pay-Offs." *Marine Ecology Progress Series* 332 (2007): 249–55.

Holling, C. S., and C. R. Allen. "Adaptive Inference for Distinguishing Credible from Incredible Patterns in Nature." *Ecosystems* 5, no. 4 (2002): 319–28.

Johannes, R. E. "The Case for Data-Less Marine Resource Management: Examples from Tropical Nearshore Fisheries." *Trends in Ecology and Evolution* 13, no. 6 (1998): 243–46.

Johnson, Kenneth G., Stephen J. Brooks, Phillip B. Fenberg, Adrian G. Glover, Karen E. James, Adrian M. Lister, Ellinor Michel, Mark Spencer, Jonathan A. Todd, Eugenia Valsami-Jones, Jeremy R. Young, and John R. Stewart. "Climate Change and Biosphere Response: Unlocking the Collections Vault." *BioScience* 61, no. 2 (2011): 147–53.

Johnson, L. C., and J. R. Matchett. "Fire and Grazing Regulate Belowground Processes in Tallgrass Prairie." *Ecology* 82, no. 12 (2001): 3377–89.

Kaplan, R., and S. Kaplan. "Adolescents and the Natural Environment: A Time Out?" In *Children and Nature: Psychological, Sociocultural, and Evolutionary Investigations*, edited by Peter H. Kahn and Stephen R. Kellert, 227–57. Cambridge, MA: MIT Press, 2002.

Kennedy, T. A., S. Naeem, K. M. Howe, J. M. H. Knops, D. Tilman, and P. Reich. "Biodiversity as a Barrier to Ecological Invasion." *Nature* 417 (2002): 636–38.

Kingsland, Sharon E. "Defining Ecology as a Science." In *Foundations in Ecology*, edited by Leslie A. Real and James H. Brown, 1–13. Chicago, Il: Univ. of Chicago Press, 1991.

Kiorboe, T., A. Andersen, V. J. Langlois, H. H. Jakobsen, and T. Bohr. "Mechanisms and Feasibility of Prey Capture in Ambush-Feeding Zooplankton." *Proceedings of the National Academy of Sciences of the United States of America* 106, no. 30 (2009): 12394–99.

Kish, Daniel. "Seeing with Sound." *New Scientist* (2009): 31–33.

Klinger, Terrie. "Address to the Western Society of Naturalists." In *Western Society of Naturalists 89th Annual Meeting*. Vancouver, BC, 2008.

Knapp, Alan K., and Charlene D'Avanzo. "Teaching with Principles: Toward More Effective Pedagogy in Ecology." *Ecosphere* 1, no. 6 (2010): art15.

Kozak, K. H., C. H. Graham, and J. J. Wiens. "Integrating GIS-Based Environmental Data into Evolutionary Biology." *Trends in Ecology & Evolution* 23, no. 3 (2008): 141–48.

Krebs. "Mapping Networks of Terrorist Cells." *Connections* 24 (2002): 43–52.

Lalak, N. 2003. Sensory Education: Stimulation of our sensory modalities increases awareness of our place in nature, fosters a sense of connection and engages us at a deeper level with our landscape and the natural world. *LandscapeAustralia* 25 (2), pp. 72-73.

Lawson, A. E. "Basic Inferences of Scientific Reasoning, Argumentation, and Discovery." *Science Education* 94, no. 2 (2010): 336–64.

Leopold, Aldo. *A Sand County Almanac with Essays on Conservation from Round River*. Oxford: Oxford Univ. Press, 1966.

Lewis, J. R. "Laboratory Charges." *Nature* 257 (1975): 640.

Lister, Adrian M. "Natural History Collections as Sources of Long-Term Datasets." *Trends in Ecology and Evolution* 26, no. 4 (2011): 153–54.

Loarie, Scott R., Lucas N. Joppa, and Stuart L. Pimm. "Satellites Miss Environmental Priorities." *Trends in Ecology & Evolution* 22, no. 12 (2007): 630–32.

Lockwood, Samuel. "Something About Crabs." *American Naturalist* 3, no. 5 (1869): 261–69.

Lopez-Medellin, X., E. Ezcurra, C. Gonzalez-Abraham, J. Hak, L. S. Santiago, and J. O. Sickman. "Oceanographic Anomalies and Sea-Level Rise Drive Mangroves Inland in the Pacific Coast of Mexico." *Journal of Vegetation Science* 22, no. 1 (2011): 143–51.

Louv, Richard. *Last Child in the Woods* Chapel Hill, NC: Algonquin Books, 2005.

Lovett, Gary M., Douglas A. Burns, Charles T. Driscoll, Jennifer C. Jenkins, Myron J. Mitchell, Lindsey Rustad, James B. Shanley, Gene E. Likens, and Richard Haeuber. "Who Needs Environmental Monitoring?" *Frontiers in Ecology and Evolution* 5, no. 5 (2007): 253–60.

Lozano-Montes, Hector M., Tony J. Pitcher, and Nigel Haggan. "Shifting Environmental and Cognitive Baselines in the Upper Gulf of California." *Frontiers in Ecology and the Environment* 6 (2008): 75–80.

MacKenzie, D. I., J. D. Nichols, J. A. Royle, K. P. Pollock, L. L. Bailey, and J. E. Hines. *Ecological-Niche Factor Analysis: How to Compute Habitat-Suitability Map without Absence Data*. San Diego, CA: Academic Press, 2006.

MacKenzie, D. I., J. D. Nichols, N. Sutton, K. Kawanishi, and L. L. Bailey. "Improving Inferences in Population Studies of Rare Species That Are Detected Imperfectly." *Ecology* 86, no. 5 (2005): 1101–13.

Martinez del Rio, Carlos. "A Natural History Curriculum for Cyborgs." In *Ecological Society of America Annual Meeting*. Albuquerque, NM, 2009.

Mazzarello, Paolo. "Museums: Stripped Assets." *Nature* 480, no. 7375 (2011): 36–38.

McBride, B. B., C. A. Brewer, M. Bricker, and M. Machura. "Training the Next Generation of Renaissance Scientists: The GK-12 Ecologists, Educators, and Schools Program at the University of Montana." *BioScience* 61, no. 6 (2011): 466–76.

McClenachan, L., J. B. C. Jackson, and M. J. H. Newman. "Conservation Implications of Historic Sea Turtle Nesting Beach Loss." *Frontiers in Ecology and the Environment* 4, no. 6 (2006): 290–96.

McDougall, K. L., J. M. Alexander, S. Haider, A. Pauchard, N. G. Walsh, and C. Kueffer. "Alien Flora of Mountains: Global Comparisons for the Development of Local Preventive Measures against Plant Invasions." *Diversity and Distributions* 17, no. 1 (2011): 103–11.

McGowan, John A. "Climate and Change in Oceanic Ecosystems; the Value of Time-Series Data." *TREE* 5, no. 9 (1990): 293–99.

McGowan, John A., Daniel R. Cayan, and LeRoy M. Dorman. "Climate-Ocean Variability and Ecosystem Response in the Northeast Pacific." *Science* 281 (1998): 210–17.

McIntire, E. J. B., and A. Fajardo. "Beyond Description: The Active and Effective Way to Infer Processes from Spatial Patterns." *Ecology* 90, no. 1 (2009): 46–56.

Menge, B. A., C. Blanchette, P. Raimondi, T. Freidenburg, S. Gaines, J. Lubchenco, D. Lohse, G. Hudson, M. Foley, and J. Pamplin. "Species Interaction Strength: Testing Model Predictions Along an Upwelling Gradient." *Ecological Monographs* 74, no. 4 (2004): 663–84.

Meyer, J. L., P. C. Frumhoff, S. P. Hamburg, and C. de la Rosa. "Above the Din but in the Fray: Environmental Scientists as Effective Advocates." *Frontiers in Ecology and the Environment* 8, no. 6 (2010): 299–305.

Millard, Candice. *The River of Doubt: Theodore Roosevelt's Darkest Journey*. New York: Broadway, 2006.

Miller, C. R., and L. P. Waits. "The History of Effective Population Size and Genetic Diversity in the Yellowstone Grizzly (Ursus Arctos): Implications for

Conservation." *Proceedings of the National Academy of Sciences* 100, no. 7 (2003): 4334–39.

Miller, Chaz. "The Garbage Barge." *Waste Age*, 1 February 2007.

Mithra, K. 2004. "Interview with Daniel Kish." *Daredevil: The Man Without Fear*. Accessed 22 April 2010, manwithoutfear.com/interviews/ddINTERVIEW.shtml?id=Kish.

Mohan, Jacqueline E., Lewis H. Ziska, William H. Schlesinger, Richard B. Thomas, Richard C. Sicher, Kate George, and James S. Clark. "Biomass and Toxicity Responses of Poison Ivy (Toxicodendron Radicans) to Elevated Atmospheric $CO_2$." *Proceedings of the National Academy of Science* 103, no. 24 (2006): 9086–89.

Molina, Mario J., and F. S. Rowland. "Stratospheric Sink for Chlorofluoromethanes: Chlorine Atom-Catalysed Destruction of Ozone." *Nature* 249 (1974): 810–12.

Moll, Remington J., Joshua J. Millspaugh, Jeff Beringer, Joel Sartwell, and Zhihai He. "A New 'View' of Ecology and Conservation through Animal-Borne Video Systems." *Trends in Ecology & Evolution* 22, no. 12 (2007): 660–68.

Montgomery, David R. *King of Fish: The Thousand-Year Run of Salmon*. Boulder, CO: Westview Press, 2003.

Morton, D. C., R. S. DeFries, J. Nagol, C. M. Souza, E. S. Kasischke, G. C. Hurtt, and R. Dubayah. "Mapping Canopy Damage from Understory Fires in Amazon Forests Using Annual Time Series of Landsat and Modis Data." *Remote Sensing of Environment* 115, no. 7 (2011): 1706–20.

Mueller, L. D. "Density-Dependent Population-Growth and Natural-Selection in Food-Limited Environments—the Drosophila Model." *American Naturalist* 132, no. 6 (1988): 786–809.

Muir, John. *Steep Trails*. Boston, MA: Houghton, 1918.

Muñoz, O., M. Montes, and T. Wilkomirsky. *Plantas Medicinales de Uso en Chile: Química y Farmacología*. Chile: Editorial Universitaria, 1999.

Mycio, Mary. *Wormwood Forest: A Natural History of Chernobyl*. Washington, DC: Joseph Henry Press, 2006.

Mysterud, Ivar, and Iver Mysterud. "Reviving the Ghost of Broad Ecology." *Journal of Social and Evolutionary Systems* 17, no. 2 (1994): 167–95.

Nemani, R. R., C. D. Keeling, H. Hashimoto, W. M. Jolly, S. C. Piper, C. J. Tucker, R. B. Myneni, and S. W. Running. "Climate-Driven Increases in Global Terrestrial Net Primary Production from 1982 to 1999." *Science* 300, no. 5625 (2003): 1560–63.

Nichols, J. D., and B. K. Williams. "Monitoring for Conservation." *Trends in Ecology & Evolution* 21, no. 12 (2006): 668–73.

Nijhuis, M. "Teaming up with Thoreau." *Smithsonian*, October 2007.
Norkko, J., S. F. Thrush, and R. M. G. Wells. "Indicators of Short-Term Growth in Bivalves: Detecting Environmental Change across Ecological Scales." *Journal of Experimental Marine Biology and Ecology* 337 (2006): 38–48.
Norment, Christopher. *Return to Warden's Grove: Science, Desire, and the Lives of Sparrows*. Iowa City, IA: Univ. of Iowa Press, 2008.
Noss, Reed. "The Naturalists Are Dying Off." *Conservation Biology* 10, no. 1 (1996): 1–3.
Nunez, M. A., and G. M. Crutsinger. "Striking a Balance between the Literature Load and Walks in the Woods." *Frontiers in Ecology and the Environment* 6, no. 3 (2008): 160–61.
O'Donohue, W., and J. A. Buchanan. "The Weaknesses of Strong Inference." *Behavior and Philosophy* 29, no. 1 (2001): 1–20.
Olson, Randy. *Don't Be Such a Scientist*. Washington, DC: Island Press, 2009.
O'Neill, Tara. "Uncovering Student Ownership in Science Learning: The Making of a Student Created Mini-Documentary." *School Science and Mathematics* 105, no. 6 (2005): 292–301.
Pain, Stephanie. "Code Red." *New Scientist*, 4 April 2009, 38–41.
Paine, Robert T. "Macroecology: Does It Ignore or Can It Encourage Further Ecological Syntheses Based on Spatially Local Experimental Manipulations?" *American Naturalist* 176, no. 4 (2010): 385–93.
Parmesan, C., and G. Yohe. "A Globally Coherent Fingerprint of Climate Change Impacts across Natural Systems." *Nature* 421, no. 6918 (2003): 37–42.
Pauchard, A., and P. B. Alaback. "Influence of Elevation, Land Use, and Landscape Context on Patterns of Alien Plant Invasions Along Roadsides in Protected Areas of South-Central Chile." *Conservation Biology* 18, no. 1 (2004): 238–48.
Pauchard, A., C. Kueffer, H. Dietz, C. C. Daehler, J. Alexander, P. J. Edwards, J. R. Arévalo, L. Cavieres, A. Guisan, S. Haider, G. Jakobs, K. McDougall, C. I. Millar, B. J. Naylor, C. G. Parks, L. J. Rew, and T. Seipel. "Ain't No Mountain High Enough: Plant Invasions Reaching High Elevations." *Frontiers in Ecology and Evolution* Online Early (2009).
Pauchard, A., M. Aguayo, E. Peña, and R. Urrutia. "Multiple Effects of Urbanization on the Biodiversity of Developing Countries: The Case of a Fast-Growing Metropolitan Area (Concepción, Chile)." *Biological Conservation* 127, no. 3 (2006): 272–81.
Pauchard, Aníbal, and Katriona Shea. "Integrating the Study of Non-Native Plant Invasions across Spatial Scales." *Biological Invasions* 8 (2006): 399–413.
Pearson, D. L., A. L. Hamilton, and T. L. Erwin. "Recovery Plan for the Endangered Taxonomy Profession." *BioScience* 61, no. 1 (2011): 58–63.

Pereira, H. M., J. Belnap, N. Brummitt, B. Collen, H. Ding, M. Gonzalez-Espinosa, R. D. Gregory, J. Honrado, R. H. G. Jongman, R. Julliard, L. McRae, V. Proenca, P. Rodrigues, M. Opige, J. P. Rodriguez, D. S. Schmeller, C. van Swaay, and C. Vieira. "Global Biodiversity Monitoring." *Frontiers in Ecology and the Environment* 8, no. 9 (2010): 458–60.

Phillips, S. J., R. P. Anderson, and R. E. Schapire. "Maximum Entropy Modeling of Species Geographic Distributions." *Ecological Modelling* 190, nos. 3–4 (2006): 231–59.

Pickett, Steward T., Clive G. Jones, and Jurek Kolasa. *Ecological Understanding: The Nature of Theory and the Theory of Nature*. Amsterdam/Boston: Elsevier/Academic Press, 2007.

Pigliucci, M. "Are Ecology and Evolutionary Biology 'Soft' Sciences?" *Annales Zoologici Fennici* 39, no. 2 (2002): 87–98.

Platt, John R. "'Strong Inference.'" *Science* 146, no. 3642 (1964): 347–53.

Poll, M., B. J. Naylor, J. M. Alexander, P. J. Edwards, and H. Dietz. "Seedling Establishment of Asteraceae Forbs Along Altitudinal Gradients: A Comparison of Transplant Experiments in the Native and Introduced Ranges." *Diversity and Distributions* 15, no. 2 (2009): 254–65.

Pollack, Andrew. "DNA Sequencing Caught in Deluge of Data." *New York Times*, 30 November 2011.

Prates-Clark, Cassia Da Conceicao, Sassan S. Saatchi, and Donat Agosti. "Predicting Geographical Distribution Models of High-Value Timber Trees in the Amazon Basin Using Remotely Sensed Data." *Ecological Modelling* (2007).

Quinn, James F., and Arthur E. Dunham. "On Hypothesis Testing in Ecology and Evolution." *American Naturalist* 122, no. 5 (1983): 620–17.

Raimondi, P. , R. Sagarin, R. Ambrose, M. George, S. Lee, D. Lohse, C. M. Miner, S. Murray, and C. Roe. "Consistent Frequency of Color Morphs in the Sea Star *Pisaster Ochraceus*." *Pacific Science* 61, no. 2 (2007): 201–10.

Reagan, Brad. "The Digital Ice Age." *Popular Mechanics*, October 2009, accessed online.

Rennie, L. J., E. Feher, Dierking. L. D., and J. H. Falk. "Toward an Agenda for Advancing Research on Science Learning in Out-of-School Settings." *Journal of Research in Science Teaching* 40, no. 2 (2003): 112–20.

Revkin, A. C. "Into the Breach." *Frontiers in Ecology and the Environment* 8, no. 6 (2010): 283–83.

Ricketts, Edward F. "Essay on Non-Teleological Thinking." In *Breaking Through: Essays, Journals and Travelogues of Edward F. Ricketts*, edited by Katharine A. Rodger, 119–33. Berkeley, CA: Univ. of California Press, 2006.

Ricketts, Edward F. "Ricketts Papers." Stanford Univ. Libraries Special Collections, 1945–1947.

Ricketts, Edward F. "The Philosophy of 'Breaking Through.'" In *Breaking Through: Essays, Journals and Travelogues of Edward F. Ricketts*, edited by Katharine A. Rodger, 89–104. Berkeley, CA: Univ. of California Press, 2006.

Ricketts, Edward F. "Transcript of Summer 1945 and 1946 Notes Based on Trips to the Outer Shores, West Coast of Vancouver Island, Queen Charlotte Islands, and So On." In *Breaking Through: Essays, Journals and Travelogues of Edward F. Ricketts*, edited by Katharine A. Rodger, 222–323. Berkeley, CA: Univ. of California Press, 2006.

Roberts, D. A., G. E. Hofmann, and G. N. Somero. "Heat-Shock Protein Expression in *Mytilus Californianus*: Acclimatization (Seasonal and Tidal-Height Comparisons) and Acclimation Effects." *Biological Bulletin* 192, no. 2 (1997): 309–20.

Rodger, Katharine A., ed. *Breaking Through: Essays, Journals and Travelogues of Edward F. Ricketts*. Berkeley, CA: Univ. of California Press, 2006.

Roemmich, Dean, and John McGowan. "Climatic Warming and the Decline of Zooplankton in the California Current." *Science* 267 (1995): 1324–26.

Root, T. L., J. T. Price, K. R. Hall, S. H. Schneider, C. Rosenzweig, and J. A. Pounds. "Fingerprints of Global Warming on Wild Animals and Plants." *Nature* 421, no. 6918 (2003): 57–60.

Roubik, David. "Competitive Interactions between Neotropical Pollinators and Africanized Honey Bees." *Science* 201 (1978): 1030–32.

Rozzi, R., J. J. Armesto, B. Goffinet, W. Buck, F. Massardo, J. Silander, M. T. Arroyo, S. Russell, C. B. Anderson, L. A. Cavieres, and J. B. Callicott. "Changing Lenses to Assess Biodiversity: Patterns of Species Richness in Sub-Antarctic Plants and Implications for Global Conservation." *Frontiers in Ecology and the Environment* 6, no. 3 (2008): 131–37.

Rundus, Aaron S., Donald H. Owings, Sanjay S. Joshi, Erin Chinn, and Nicholas Giannini. "Ground Squirrels Use an Infrared Signal to Deter Rattlesnake Predation." *Proceedings of the National Academy of Science* 104, no. 36 (2007): 14372–76.

Saenz-Arroyo, Andrea, Callum M. Roberts, Jorge Torre, Micheline Carino-Olvera, and Roberto R. Enriquez-Andrade. "Rapidly Shifting Environmental Baselines among Fishers of the Gulf of California." *Proceedings of the Royal Society B* 272 (2005): 1957–62.

Sagarin, R. D. "Historical Studies of Species' Responses to Climate Change: Promises and Pitfalls." In *Wildlife Responses to Climate Change: North American*

*Case Studies*, edited by Stephen H. Schneider and Terry L. Root: Island Press, 2001.

Sagarin, R. D. "Natural Security for a Variable and Risk-Filled World." *Homeland Security Affairs* 6, no. 3 (2010).

Sagarin, R. D. "Phenology—False Estimates of the Advance of Spring." *Nature* 414, no. 6864 (2001): 600.

Sagarin, R. D. "Science Communication: More Than Words." *Frontiers in Ecology and the Environment* 9, no. 8 (2010): 458.

Sagarin, R. D. *Learning from the Octopus: How Secrets of Nature Can Help Us Fight Terrorist Attacks, Natural Disasters, and Disease.* New York. Basic Books (2012).

Sagarin, R. D., and A. Pauchard. "Observational Approaches in Ecology Open New Ground in a Changing World " *Frontiers in Ecology and the Environment* 8 (2010): 379–86.

Sagarin, R. D., and F. Micheli. "Climate Change in Nontraditional Data Sets." *Science* 294 (2001): 811.

Sagarin, R. D., and George N. Somero. "Complex Patterns of Heat-Shock Protein 70 Expression across the Southern Biogeographic Ranges of the Intertidal Mussel *Mytilus Californianus* and Snail *Nucella Ostrina*." *Journal of Biogeography* 33, no. 4 (2006): 622–30.

Sagarin, R. D., and S. D. Gaines. "Geographical Abundance Distributions of Coastal Invertebrates: Using One-Dimensional Ranges to Test Biogeographic Hypotheses." *Journal of Biogeography* 29, no. 8 (2002a): 985–97.

Sagarin, R. D., and S. D. Gaines. "The 'Abundant Centre' Distribution: To What Extent Is It a Biogeographic Rule?" *Ecology Letters* 5 (2002b): 137–47.

Sagarin, R. D., C. S. Alcorta, S. Atran, D. T. Blumstein, G. P. Dietl, M. E. Hochberg, D. D. P. Johnson, S. Levin, E. M. P. Madin, J. S. Madin, E. M. Prescott, R. Sosis, T. Taylor, J. Tooby, and G. J. Vermeij. "Decentralize, Adapt, and Cooperate." *Nature* 465, no. 7296 (2010): 292–93.

Sagarin, R. D., J. Carlsson, M. Duval, W. Freshwater, M. H. Godfrey, W. Litaker, R. Munoz, R. Noble, T. Schultz, and B. Wynne. "Bringing Molecular Tools into Environmental Resource Management: Untangling the Molecules to Policy Pathway." *PLoS Biology* 7, no. 3 (2009): 426–30.

Sagarin, R. D., R. Ambrose, B. Becker, J. Engle, J. Kido, S. Lee, C. M. Miner, S. Murray, P. Raimondi, D. Richards, and C. Roe. "Ecological Impacts on the Limpet Lottia Gigantea Populations: Human Pressure over a Broad Scale on Island and Mainland Intertidal Zones." *Marine Biology* 150, no. 3 (2007): 399–413.

Sagarin, R. D., Steven D. Gaines, and Brian Gaylord. "Moving beyond Assumptions to Understand Abundance Distributions across the Ranges of Species." *Trends in Ecology & Evolution* 21, no. 9 (2006): 524–30.

Sagarin, R. D., W. F. Gilly, C. H. Baxter, N. Burnett, and J. Christensen. "Remembering the Gulf: Changes to the Marine Communities of the Sea of Cortez since the Steinbeck and Ricketts Expedition of 1940." *Frontiers in Ecology and the Environment* 6, no. 7 (2008): 374–81.

Sageman, Marc. *Understanding Terror Networks*. Philadelphia, PA: Univ. of Pennsylvania Press, 2004.

Salomon, Anne K., Nick M. Tanape Sr., and Henry P. Huntington. "Serial Depletion of Marine Invertebrates Leads to the Decline of a Strongly Interacting Grazer." *Ecological Applications* 17, no. 6 (2007): 1752–70.

Schaller, George B. "The Pleasure of Observing." In *Field Notes on Science and Nature*, edited by Michael R. Canfield, 19–32. Cambridge, MA: Harvard Univ. Press, 2011.

Schmitt, Russell J., and Craig W. Osenberg, eds. *Detecting Ecological Impacts: Concepts and Applications in Coastal Habitats*. San Diego, CA: Academic Press, 1996.

Schwartz-Shea, Peregrine. "Contribution to a Review Symposium: Beyond the Tragedy of the Commons." *Perspectives on Politics* 8, no. 2 (2010): 587–90.

Scott, Anthony. "Trust Law, Sustainability, and Responsible Action." *Ecological Economics* 31 (1999): 139–54.

Seastedt, T. R., R. J. Hobbs, and K. N. Suding. "Management of Novel Ecosystems: Are Novel Approaches Required?" *Frontiers in Ecology and the Environment* 6, no. 10 (2008): 547–53.

Seipel, Tim, Christoph Kueffer, Lisa J. Rew, Curtis C. Daehler, Aníbal Pauchard, Bridgett J. Naylor, Jake M. Alexander, Peter J. Edwards, Catherine G. Parks, José Ramon Arevalo, Lohengrin A. Cavieres, Hansjörg Dietz, Gabi Jakobs, Keith McDougall, Rüdiger Otto, and Neville Walsh. "Processes at Multiple Scales Affect Richness and Similarity of Non-Native Plant Species in Mountains around the World." *Global Ecology and Biogeography* (2011).

Shimbun, Asashi. "Fisherman Powered out to Meet Giant Tsunami." *New York Times*, 17 March 2011, accessed online.

Simberloff, D. "Competition Theory, Hypothesis-Testing, and Other Community Ecological Buzzwords." *American Naturalist* 122, no. 5 (1983): 626–35.

Simberloff, Daniel. "Community Ecology: Is It Time to Move On?" *American Naturalist* 163, no. 6 (2004): 787–99.

Skole, D., and C. Tucker. "Tropical Deforestation and Habitat Fragmentation in the Amazon—Satellite Data from 1978 to 1988." *Science* 260, no. 5116 (1993): 1905–10.

Sobel, D. *Beyond Ecophobia: Reclaiming the Heart of Nature Education*. 45 vols. Great Barrington, MA: Orion Society, 1996.

Somero, G. 1995. Proteins and temperature. *Annual Review of Physiology* 57: 43-68.

Southward, A. J., S. J. Hawkins, and M. T. Burrows. "Seventy Years' Observations of Changes in Distribution and Abundance of Zooplankton and Intertidal Organisms in the Western English Channel in Relation to Rising Sea Temperature." *Journal of Thermal Biology* 20, no. 1–2 (1995): 127–55.

Steinbeck, John, and Edward F. Ricketts. *Sea of Cortez: A Leisurely Journal of Travel and Research, with a Scientific Appendix Comprising Materials for a Source Book on the Marine Animals of the Panamic Faunal Province.* Mamaroneck, NY: P. P. Appel, 1941, reprinted 1971.

Stockwell, D. R. B. , and D. P. Peters. "The GARP Modelling System: Problems and Solutions to Automated Spatial Prediction." *International Journal of Geographical Information Systems* 13 (1999): 143–58.

Stohlgren, T. J., C. Flather, C. S. Jarnevich, D. T. Barnett, and J. Kartesz. "Rejoinder to Harrison (2008): The Myth of Plant Species Saturation." *Ecology Letters* 11, no. 4 (2008): 324–26.

Stohlgren, T. J., D. Binkley, G. W. Chong, M. A. Kalkhan, L. D. Schell, K. A. Bull, Y. Otsuki, G. Newman, M. Bashkin, and Y. Son. "Exotic Plant Species Invade Hot Spots of Native Plant Diversity." *Ecological Monographs* 69, no. 1 (1999): 25–46.

Stohlgren, T. J., D. T. Barnett, and J. Kartesz. "The Rich Get Richer: Patterns of Plant Invasions in the United States." *Frontiers in Ecology and the Environment* 1, no. 1 (2003): 11–14.

Stohlgren, T. J., K. A. Bull, Y. Otsuki, C. A. Villa, and M. Lee. "Riparian Zones as Havens for Exotic Plant Species in the Central Grasslands." *Plant Ecology* 138, no. 1 (1998): 113–25.

Stohlgren T. J., P. Pysek, J. Kartesz, M. Nishino, A. Pauchard, M. Winter, J. Pino, D. M. Richardson, J. R. U. Wilson, B. R. Murray, M. L. Phillips, L. Ming-yang, L. Celesti-Grapow, X. Font. 2011. "Widespread plant species: natives versus aliens in our changing world." *Biological invasions* 13:1931–1944.

Stott, Rebecca. *Darwin and the Barnacle: The Story of One Tiny Creature and History's Most Spectacular Scientific Breakthrough.* New York: W. W. Norton, 2003.

Suter, G. W. "Abuse of Hypothesis Testing Statistics in Ecological Risk Assessment." *Human and Ecological Risk Assessment* 2, no. 2 (1996): 331–47.

Taleb, Nassim Nicholas. *The Black Swan: The Impact of the Highly Improbable.* New York: Random House, 2007.

Tamm, Eric Enno. *Beyond the Outer Shores: The Untold Odyssey of Ed Ricketts, the Pioneering Ecologist Who Inspired John Steinbeck and Joseph Campbell.* New York: Thunder's Mouth Press, 2004.

Tang, J. M., L. Wang, and Z. J. Yao. "Analyzing Urban Sprawl Spatial Fragmentation Using Multi-Temporal Satellite Images." *GIScience & Remote Sensing* 43, no. 3 (2006): 218–32.

Taylor, A. F. , and F. E. Kuo. "Is Contact with Nature Important for Healthy Child Development? State of the Evidence." In *Children and Their Environments: Learning, Using, and Designing Spaces*, edited by Christopher Spencer and Mark Blades, 121–40. Cambridge, UK ; New York: Cambridge Univ. Press, 2006.

Taylor, A. F., F. E. Kuo, and W. C. Sullivan. "Coping with ADD—The Surprising Connection to Green Play Settings." *Environment and Behavior* 33, no. 1 (2001): 54–77.

Thomashow, C. "Adolescents and Ecological Integrity." In *Children and Nature: Psychological, Sociocultural, and Evolutionary Investigations*, edited by Peter H. Kahn and Stephen R. Kellert, 259–78. Cambridge, MA: MIT Press, 2002.

Tilman, D. "The Ecological Consequences of Changes in Biodiversity: A Search for General Principles." *Ecology* 80, no. 5 (1999): 1455–74.

Tollefson, Jeff. "US Launches Eco-Network." *Nature* 476 (2011): 135.

Tomppo, Erkki. *National Forest Inventories: Pathways for Common Reporting*. Heidelberg: Springer, 2009.

Travis, D. J., A. M. Carleton, and R. G. Lauritsen. "Climatology: Contrails Reduce Daily Temperature Range—A Brief Interval When the Skies Were Clear of Jets Unmasked an Effect on Climate." *Nature* 418, no. 6898 (2002): 601–01.

Turco, R. P., O. B. Toon, T. P. Ackerman, J. B. Pollack, and C. Sagan. "Nuclear Winter—Global Consequences of Multiple Nuclear-Explosions." *Science* 222, no. 4630 (1983): 1283–92.

Turner, W., S. Spector, N. Gardiner, M. Fladeland, E. Sterling, and M. Steininger. "Remote Sensing for Biodiversity Science and Conservation." *Trends in Ecology & Evolution* 18, no. 6 (2003): 306–14.

Turnipseed, M., S. E. Roady, R. Sagarin, and L. B. Crowder. "The Silver Anniversary of the United States' Exclusive Economic Zone: Twenty-Five Years of Ocean Use and Abuse, and the Possibility of a Blue Water Public Trust Doctrine." *Ecology Law Quarterly* 36, no. 1 (2009): 1–70.

Underwood, E. C., S. L. Ustin, and C. M. Ramirez. "A Comparison of Spatial and Spectral Image Resolution for Mapping Invasive Plants in Coastal California." *Environmental Management* 39, no. 1 (2007): 63–83.

Vadas, R. L. "The Anatomy of an Ecological Controversy—Honeybee Searching Behavior." *Oikos* 69, no. 1 (1994): 158–66.

Valone, T. J. "Are Animals Capable of Bayesian Updating? An Empirical Review." *Oikos* 112, no. 2 (2006): 252–59.

Van Valkenburgh, Blaire, Xiaoming Wang, and John Damuth. "Cope's Rule, Hypercarnivory, and Extinction in North American Canids." *Science* 306 (2004): 101–04.

Verchot, L. V., P. M. Groffman, and D. A. Frank. "Landscape Versus Ungulate Control of Gross Mineralization and Gross Nitrification in Semi-Arid Grasslands of Yellowstone National Park." *Soil Biology & Biochemistry* 34, no. 11 (2002): 1691–99.

Vermeij, Geerat. "Science, Blindness, and Evolution: The Common Theme Is Opportunity." *Journal of Science Education for Students with Disabilities* 10, no. 1 (2002): 1–3.

Vermeij, Geerat. *Nature: An Economic History*. Princeton: Princeton Univ. Press, 2004.

Vierling, K. T., L. A. Vierling, W. A. Gould, S. Martinuzzi, and R. M. Clawges. "Lidar: Shedding New Light on Habitat Characterization and Modeling." *Frontiers in Ecology and the Environment* 6, no. 2 (2008): 90–98.

Vierling, L. A., S. Martinuzzi, G. P. Asner, J. Stoker, and B. R. Johnson. "Lidar: Providing Structure." *Frontiers in Ecology and the Environment* 9, no. 5 (2011): 261–62.

Vitousek, Peter M., Harold A. Mooney, Jane Lubchenco, and Jerry M. Melillo. "Human Domination of Earth's Ecosystems." *Science* 277 (1997): 494–99.

Wade, Paul R. "Bayesian Methods in Conservation Biology." *Conservation Biology* 14, no. 5 (2000): 1308–16.

Walker, Thomas C. "The Perils of Paradigm Mentalities: Revisiting Kuhn, Lakatos, and Popper." *Perspectives on Politics* 8, no. 2 (2010): 433–51.

Wandeler, Peter, Paquita E. A. Hoeck, and Lukas F. Keller. "Back to the Future: Museum Specimens in Population Genetics." *Trends in Ecology & Evolution* 22, no. 12 (2007): 635–42.

Webb, Robert H., Diane E. Boyer, and Raymond M. Turner, eds. *Repeat Photography: Methods and Applications in the Natural Sciences*. Washington, DC: Island Press, 2010.

Weber, T. P. "A Plea for a Diversity of Scientific Styles in Ecology." *Oikos* 84, no. 3 (1999): 526–29.

Weiner, Jacob. "On the Practice of Ecology." *Journal of Ecology* 83 (1995): 153–58.

Wenner, A. M. "Concept-Centered Versus Organism-Centered Biology." *American Zoologist* 29, no. 3 (1989): 1177–97.

Whitmer, A., L. Ogden, J. Lawton, P. Sturner, P. M. Groffman, L. Schneider, D. Hart, B. Halpern, W. Schlesinger, S. Raciti, N. Bettez, S. Ortega, L. Rustad, S. T. A. Pickett, and M. Killelea. "The Engaged University: Providing a Platform for Research That Transforms Society." *Frontiers in Ecology and the Environment* 8, no. 6 (2010): 314–21.

Williams, J. W., and S. T. Jackson. "Novel Climates, No-Analog Communities, and Ecological Surprises." *Frontiers in Ecology and the Environment* 5, no. 9 (2007): 475–82.

Wilson, E. O. *Naturalist*. New York: Warner Books, 1994.

Wilson, Edward O. *Consilience: The Unity of Knowledge*. New York: Alfred A. Knopf, 1998.

# ABOUT THE AUTHORS

**RAFE SAGARIN** is a marine ecologist and environmental policy analyst at the University of Arizona. He uses natural history observations and historical data sets from writers, naturalists, artists, and even gamblers to reassemble historical patterns of ecosystem change, including changes to the Gulf of California since the 1940 expedition of John Steinbeck and Ed "Doc" Ricketts. Using an approach inspired by the ecological philosophy of Ricketts, Rafe applies basic observations of nature to issues of broad societal interest, including conservation biology, protecting public-trust resources, and making our responses to terrorism and other security threats more adaptable.

Rafe is a recipient of a Guggenheim Fellowship and was a Geological Society of America Congressional Science Fellow in the office of U.S. Representative (and later U.S. Secretary of Labor) Hilda Solis. He has taught ecology and environmental policy at the University of Arizona, Duke University, California State University–Monterey Bay, Stanford University, and the University of California–Los Angeles. His research has appeared in *Science, Nature, Conservation Biology, Ecological Monographs, Trends in Ecology and Evolution, Foreign Policy, Homeland Security Affairs,* and other leading journals, magazines, and newspapers. With Terence Taylor, he is the editor of the volume *Natural Security: A Darwinian Approach to a Dangerous World* (University of California Press, 2008) and the author of *Learning from the Octopus: How Secrets from Nature Can Help Us Fight Terrorist Attacks, Natural Disasters, and Disease* (Basic Books, 2012).

**ANÍBAL PAUCHARD** was born in Santiago, Chile, in 1974. He has a bachelor's degree in forestry from the Universidad de Concepción (1998) and a PhD from the College of Forestry and Conservation at the University of Montana (2002). Since 2003 he has worked at the Faculty of Forest Sciences at the Universidad de Concepción, where he is currently associate professor and director of the Laboratorio de Invasiones Biológicas (LIB).

He is also adjunct researcher at the Institute of Ecology and Biodiversity (IEB, Chile). He is currently adjunct professor at both the University of Montana and North Carolina State University. Since 2010, he has been the head of the undergraduate program of Natural Resource Conservation at the Universidad de Concepción.

His research is focused mainly on the ecology and biogeography of biological invasions and their impacts on biodiversity and ecosystem functions. He has been studying alien-plant invasions in natural and seminatural areas across elevational gradients using multi-scale approaches. Along with researchers from several countries, he co-founded the Mountain Invasion Research Network (MIREN) to search for the causes and impacts of invasion processes in mountain environments. Additionally, he has been looking at larger intercontinental scales, specifically comparing plant invasions between California and Chile (two regions with similar climates), and now between Chile and New Zealand.

Aníbal is also becoming increasingly interested in broader issues in ecology and the management of natural resources such as ecology education, conservation psychology, and differences between developed and developing countries in how they deal with conservation problems.

# ABOUT THE CONTRIBUTORS

**PAUL DAYTON** is a professor at the Scripps Institute of Oceanography. He is a marine ecologist and naturalist, and a recipient of the Lifetime Achievement Award from the Western Society of Naturalists.

**BRETT DICKSON** is an assistant research professor at the School of Earth Sciences and Environmental Sustainability, Northern Arizona University. His work applies tools from landscape and wildlife ecology to solve conservation problems throughout North America.

**DAVID DOBKIN** is executive director of the Greater Hart-Sheldon Conservation Fund and the High Desert Ecological Research Institute in Bend, Oregon. He has conducted research on a wide variety of taxa in arid landscapes of western North America, focusing in particular on the ecology of shrub-steppe landscapes, with an emphasis on riparian bird communities.

**THOMAS L. FLEISCHNER** is a professor at Prescott College. His work is always rooted in natural history, ecology, and conservation biology, but plies the terrain at the margins of disciplines. He is the founding president of the Natural History Network (naturalhistorynetwork.org) and editor of *The Way of Natural History*, an anthology.

**ERICA FLEISHMAN** is a researcher at the John Muir Institute of the Environment, University of California–Davis. Her research focuses on the integration of conservation science with the management of public and private lands in the western United States.

**BRENDON LARSON** is an associate professor in the Faculty of Environment, University of Waterloo. He is an interdisciplinary social scientist who integrates his lifelong experience as a naturalist and a biologist with current research on the social dimensions of biodiversity conservation.

**JULIE LOCKWOOD** is an associate professor in the Department of Ecology, Evolution, and Natural Resources at Rutgers University. Her research is a cross-section of conservation biology, biogeography, and invasion ecology. She has a very active and diverse laboratory, and she has co-written two books (*Avian Invasions* and *Invasion Ecology*), and co-edited one more (*Biotic Homogenization*).

**GARY NABHAN** holds the W. K. Kellogg Endowed Chair in Southwest Borderlands Food and Water Security at the University of Arizona's Southwest Center. He is an internationally celebrated nature writer, ethnobotanist, conservation biologist, and sustainable agriculture activist who has been called "the father of the local-food movement" by *Utne Reader* and *Mother Earth News*.

**BARRON ORR** is an assistant professor in the Office of Arid Lands Studies at the University of Arizona. His international activities focus on the development of a formative and participatory approach to environmental evaluation.

**STUART L. PIMM** is the Doris Duke Professor of Conservation Ecology in the Nicholas School of the Environment at Duke University. His interests include environmental sciences, policy, and conservation. Pimm's experience lies in species extinctions and what can be done to prevent them, as well as the loss of tropical forests and its consequences to biodiversity.

**KIRSTEN ROWELL** is a curator of malacology at the Burke Museum of Natural History and Culture and an acting assistant professor in the Department of Biology at the University of Washington. Her research reaches across disciplines (geology, ecology, and conservation biology), and she uses ancient skeletal remains to reveal the untold stories of fish and clams that have lived through large-scale human alterations to their habitats.

**RICARDO ROZZI** the director of the Subantarctic Biocultural Conservation Program coordinated by the Universidad de Magallanes (UMAG) and the Institute of Ecology and Biodiversity in Chile, and the University of North Texas (UNT) in the United States. He is a professor at UNT, and his research integrates ecological sciences and environmental philosophy into biocultural conservation.

**ANNE SALOMON** is a marine ecologist and assistant professor at Simon Fraser University. She seeks to inform ecosystem approaches to marine conservation by advancing our understanding of how human disturbances alter the productivity, biodiversity, and resilience of marine food webs.

**STEVEN SESNIE** is a spatial ecologist with the U.S. Fish and Wildlife Service, Southwest Region, in Albuquerque, New Mexico. He specializes in remote sensing, wildlife habitat characterization, and forest ecology.

**TOM STOHLGREN** is a research ecologist with the U.S. Geological Survey and senior research scientist at the National Resource Ecology Laboratory at Colorado State University. He is interested in human-altered ecosystems and the local-to-global effects of alien plants, animals, and diseases.

**GEERAT J. VERMEIJ** is a distinguished professor in the Department of Geology at the University of California–Davis. He is an expert on marine ecology and paleoecology. He has studied the functional morphology of marine mollusks and the coevolutionary reactions between predators and prey, and their effects on morphology, ecology, and evolution.

**JAKE F. WELTZIN** is an ecologist with the U.S. Geological Survey and the executive director of the USA National Phenology Network. His interests encompass how the structure and function of plant communities and ecosystems might respond to global environmental change, including atmospheric chemistry, climate change, and biological invasions.

**KRISTIN WISNESKI** is a master's degree candidate in rangeland ecology and management in the School of Natural Resources and the Environment at the University of Arizona. Her research focuses on the potential that technology holds to help young people learn about science while solving problems in their communities and the environment.

# INDEX

Figures/photos/illustrations are indicated by an "f" and tables by a "t."